PRACTICAL SPEECH
USER INTERFACE
DESIGN

Human Factors and Ergonomics

Series Editor

Gavriel Salvendy

Professor Emeritus
School of Industrial Engineering
Purdue University

Chair Professor & Head
Dept. of Industrial Engineering
Tsinghua Univ., P.R. China

PRACTICAL SPEECH USER INTERFACE DESIGN

James R. Lewis

CRC Press
Taylor & Francis Group
Boca Raton   London   New York

CRC Press is an imprint of the
Taylor & Francis Group, an **informa** business

CRC Press
Taylor & Francis Group
6000 Broken Sound Parkway NW, Suite 300
Boca Raton, FL 33487-2742

© 2011 by Taylor and Francis Group, LLC
CRC Press is an imprint of Taylor & Francis Group, an Informa business

No claim to original U.S. Government works

Printed in the United States of America on acid-free paper
10 9 8 7 6 5 4 3 2 1

International Standard Book Number: 978-1-4398-1584-7 (Hardback)

Visit the Taylor & Francis Web site at
http://www.taylorandfrancis.com

and the CRC Press Web site at
http://www.crcpress.com

Contents

Foreword

This book is being published at an opportune time, since in the second decade of the 21st century, speech user interfaces seem to be everywhere. You need service from a company, you call, and the next thing you know you're talking to an automated system. In 2001, according to a Pew Charitable Trust survey, 94% of Americans found it very frustrating to call a business and then have to deal with recorded messages—a sentiment with which many of us would sympathize, even ten years later. When parodies of interactive voice response systems show up on National Public Radio ("The Complexities of Modern Love in the Digital Age," 2002) and Saturday Night Live ("Julie the Operator Lady," 2006), and Web sites dedicated to bypassing such systems become popular (http://gethuman.com/), you don't just have a trend, you have a phenomenon. This book presents the culmination of extensive experience by the author in research, innovation (holder of over 70 patents), and design of products with speech interfaces which are used worldwide. The book presents comprehensive coverage of speech technologies including speech recognition and production, and ten key concepts in human language and communication. The book concludes with a description of how to build effective, efficient, and pleasant interactive voice response applications. This book is invaluable for researchers and practitioners in computer-based voice communication.

Gavriel Salvendy
Series Editor
www.ie.tsinghua.edu.cn/Salvendy

Preface

If speech is the most natural form of communication, then why do we often find it so hard to use speech to communicate with machines? In short, machines aren't human, so using speech to communicate with them is anything but natural. The techniques for designing usable speech user interfaces are not obvious, and must be informed by a combination of critically interpreted scientific research and leading design practices. This book draws upon the key scientific disciplines of psychology, human–computer interaction, human factors, linguistics, communication theory, and service science, the artistic disciplines of auditory design and script writing, and experience in the assessment, design, and deployment of real-world speech applications. The goal is a comprehensive yet concise survey of practical speech user interface design for interactive voice response (IVR) applications.

Although several books are available on this topic, the two that, in my opinion, have provided good research- and practice-based guidance for the design of speech-enabled interactive voice response applications are Balentine & Morgan's (2001) *How to Build a Speech Recognition Application* and Cohen, Giangola, and Balogh's (2004) *Voice User Interface Design*. There has, however, been significant research published in the design of speech user interfaces over the past 10 years that has led to the modification of some of the design practices taught in those books. In some cases I've had the opportunity to participate in this research and, due to my consulting work, have also confronted the problem of translating this research into design practice. It's often challenging, but also unbelievably rewarding, to start from a high-level conceptualization of an IVR system, work out the design details, get the system built and tested, and then listen to callers successfully using the system.

This book focuses on the design of speech user interfaces for IVR applications. There are many other applications for speech technologies, such as dictation systems, command and control, and multimodal applications, that are outside the scope of this book and for which I do not know of any recent practical book-length treatments. This is probably because IVR applications involve significant investments from the enterprises that use them, so this is where a significant amount of current applied speech development activity has occurred. Also, there's no point in working out a detailed user interface for an application only to find that no one can build it in the foreseeable future—thus this book's focus on practical speech user interface design for IVR applications.

User interfaces for IVR applications are especially challenging because they require human–computer interaction solely through the audio channel without any visual display. Furthermore, most IVR applications must be immediately usable because callers often have little experience with the application and have no alternative means for learning how to use the application. Whether the IVR will route the call to an agent or provide self-service, it must efficiently guide the caller through the task or risk losing the call (and possibly a customer).

If you're a speech user interface designer, then you can compare the practices taught in this book with your own. With any luck, you'll come across some new ways of approaching speech user interface design that will sharpen your skills, whether or not you agree with everything. If you manage speech user interface design projects, then you should be able to use this information to get a better understanding of what your designers do and how to avoid some unfortunately common but ineffective design practices. For students, I've tried to provide the references that support the design practices put forth in this book so you can work your way back through the supporting scientific literature. Also, where there are as yet insufficient published studies to address certain issues, I've tried to point this out in the hope that those of you who are graduate students might find a research topic that grabs your interest, and the pursuit of which will add to the body of research that guides speech user interface design.

Acknowledgments

As I approach my 30th year as an IBM human factors engineer, I realize I owe much to IBM. From the beginning, through my set of excellent managers, IBM fostered a research and development environment that encouraged participation in the broader human factors community, including the publication of research papers. From 1988–1989 I had the opportunity to work at the T.J. Watson Research Center and to take classes in human–computer interaction and applied statistics at Teachers College, Columbia University. From 1993 to 1996, IBM supported the pursuit of my doctorate in psycholinguistics at Florida Atlantic University (FAU). Most recently IBM has been very generous in granting me time to complete this book. Even if you work in just one area of IBM, you're sure to have many managers over the course of three decades, and I have had many excellent managers. I especially want to thank Sheri Daye, Dana Vizner, Rocco Palermo, Kim Kemble, Don Davis, and Bill Hrapchak. I also want to thank Pete Kennedy, who showed me the human factors ropes my first day at IBM and has remained a great friend since that day.

My primary academic focus has been in experimental psychology, first in engineering psychology at New Mexico State University (NMSU) in the 1970s then in psycholinguistics at FAU in the 1990s. At FAU I had the opportunity to work with Lew Shapiro on syntactic and semantic psycholinguistic research, and with Betty Tuller on the mechanics of speech. At NMSU I studied with many talented experimental psychologists who also had an interest in the emerging field of human–computer interaction, in particular Roger Schvaneveldt and Ken Paap. Of particular value to an experimental psychologist is training in statistical thinking and critical evaluation of research—for both of these I am deeply indebted to James V. Bradley.

I have worked with many outstanding graduate students who served as human factors interns at IBM. There's nothing like working with graduate students to act as a continual reminder of how much more there is to learn about the design and evaluation of systems intended for human use and for fascinating intellectual discussion. In particular I want to thank Wallace Sadowski, Melanie Polkosky, Patrick Commarford, Cheryl Kotan, and Barbara Millet for the times we spent together planning, executing, and discussing research related to speech user interface design.

I also want to express my appreciation to the people in my department at IBM with whom I work on a daily basis at the task of helping our customers improve their IVR applications. Our team includes project managers, architects, human factors specialists, and software engineers. With good humor and great intelligence they keep the job interesting and entertaining. Thanks to Stephen Hanley, Helene Lunden, Peeyush Jaiswal, Pradeep Mansey, Melanie

Polkosky, Chip Agapi, Alan McDonley, Michael Mirt, Val Peco, Greg Purdy, Sibyl Sullivan, Fang Wang, and Frankie Wilson.

Thanks to Gavriel Salvendy of Purdue University for the many opportunities he has presented to me over the years, including the invitation to write this book for his Human Factors and Ergonomics Series at CRC Press/Taylor & Francis, and to Sarah Wayland, currently at the University of Maryland, for encouraging me to take this on. Many thanks also to Patrick Commarford, Wallace Sadowski, Barbara Millet, Bruce Balentine, John Chin, Melanie Polkosky, and Peeyush Jaiswal for their thoughtful reviews of the first draft of this book. It's a cliché, but a true cliché, that I have a debt to the reviewers for all that is right with the book, but I bear full responsibility for any shortcomings.

I owe much to those who have published research or have written about practice in speech user interface design. In keeping with my training as an experimental psychologist, I have read much of this work critically in the preparation of this book, and in some cases have disagreed with the researchers, either in their methods or interpretation of their results. Whether I have agreed or disagreed, however, I have great respect for those who design experiments, carry them out, and publish their results. This scientific work is difficult but essential, and without their efforts to provide empirical data to guide design, design would be little but intuition at best; at worst, it would be guesswork. The design of any reasonably complex system will require human design decisions that go beyond the available research, but designers all benefit by making those decisions guided, to as great an extent as possible, by published experimentation.

This is my first book, and I had no idea how much of my time it would take to finish it. I thank my wife, Cathy, for her patience during this time, and for the loving support she is always ready to give. I am a lucky man.

The Author

 **James R. (Jim) Lewis, PhD,** is a senior human factors engineer (at IBM since 1981), with a primary focus on the design and evaluation of speech applications since 1993. He is a certified human factors professional with a PhD in experimental psychology (psycholinguistics), an MA in engineering psychology, and an MM in music theory and composition. Jim is an internationally recognized expert in usability testing and measurement, contributing (by invitation) the chapter on usability testing for the third edition of the *Handbook of Human Factors and Ergonomics* and presenting tutorials on usability testing and metrics at various professional conferences. He was the technical team lead for the human factors/usability group working in IBM speech product development from 1999 through 2005 and has experience in all areas of speech system usability (including desktop systems, embedded systems, text-to-speech systems, speech interactive voice response applications, and natural language understanding technologies).

Jim is the author of IBM's published guidelines for speech user interface design, and has consulted since 2001 on speech projects in the United States and internationally for major customers in diverse areas such as consumer electronics, 401(k) plan management, insurance, reservations, telecom, banking, emergency road service, entertainment, state benefits, human resources, and telematics. Jim is an IBM master inventor with 70 patents issued to date by the U.S. Patent Office (142 worldwide). He has published 19 journal articles, 5 book chapters, 43 professional conference papers, and more than 160 IBM technical reports. He currently serves on the editorial boards of the *International Journal of Human–Computer Interaction* and the *Journal of Usability Studies*, and is on the scientific advisory board of the Center for Research and Education on Aging and Technology Enhancement (CREATE). He is a member of the Usability Professionals Association, the Human Factors and Ergonomics Society, and the American Psychological Association, and is a fourth-degree black belt and certified instructor with the American Taekwondo Association.

1

Introduction

Foundations of Speech User Interface Design

The design of a speech user interface (SUI, also called voice user interface, or VUI) is not easy. As shown in Figure 1.1, effective SUI design for interactive voice response (IVR) applications—the applications that enterprises use to provide phone-based customer service—rests on three major foundational bodies of knowledge. Without speech technologies, of course, there would be no need for SUI design. Human factors engineering provides the philosophy and methods to design with an understanding of the interactions between humans and the other elements of a system, making systems compatible with the needs, abilities, and limitations of people (Lewis, 2011; Wickens, Lee, Liu, & Gordon-Becker, 2004). Of particular importance to the design of systems that provide customer service is the emerging discipline of service science (Spohrer & Maglio, 2008) and especially its subdiscipline of market research in the characteristics of successful self-service technologies (Bitner, Ostrom, & Meuter, 2002).

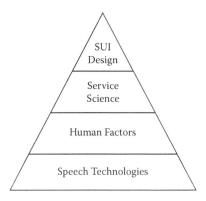

FIGURE 1.1
Foundations of speech user interface design for interactive voice responses.

A Focus on Research-Based Design Guidance

There has been a growing call in the SUI design community for more applied research and data-driven design (Bloom, Tranter, Hura, Krogh, McKienzie, Parks, & Tucker, 2008; Hura, 2009; Pieraccini & Hunter, 2010; Polkosky, 2008; Suhm, 2008). For example:

- "Without research, VUI design advances by trial and error." (Hura, 2009, p. 7)
- "When IVR design methods yield different plausible designs, it is often impossible to decide which design works best just by applying guidelines without empirical evaluation." (Suhm, 2008, p. 3)
- "Voice interaction designers, the authors included, espouse best practices like they espouse carbon dioxide. The problem is, do we know for sure that they are true? How do we know that? What data do we have to back it up? If we do have data, is it sufficient to back up our claim? And let's say for argument's sake that for any particular one we actually do have data. There are plenty more claims out there that are backed up more by anecdotal evidence and gut instinct." (Bloom et al., 2008, p. 1)
- "So, art and science are companions in the pursuit of excellence and desired effect, complementing and even relying on each other. However, while VUI producers have long used the tools and techniques of the trade, rarely has there been a full use of science behind voice interaction and an application of what can be learned from the responses provoked by the voice system. This ignorance of the data and its meaning has been a shortcoming and even a failure point for many speech deployments." (Pieraccini & Hunter, 2010, pp. 57–58)

In keeping with the philosophy and traditions of human factors engineering, this book has a strong focus on research-based design guidance. As Meister (1999, p. 75) pointed out, "the most fundamental problem of system development is the translation of behavioral principles into design characteristics." When crafting a specific design, it is not possible to have research-based guidance for every decision, but it is important to use such guidance in the design situations for which it is available. As part of the preparation for this book, I conducted numerous critical literature reviews and have provided, using the standard APA citation style, pointers to the research literature on SUI IVR design and related topics.

To help articulate the rationale behind various SUI design guidelines, the book includes a number of detailed discussions of the applicable research. When abstracting design guidance from research papers, it is inadequate to just say, "Research supports this guideline." There can be devils in the details, so for key research papers the discussions include information about the participants, their specific tasks (including, when appropriate, the exact wording of prompts and messages), the experimental outcomes, and an interpretation of those outcomes. This can be of value to SUI designers when determining the applicability of the research to the design problem at hand, and when explaining design choices to other stakeholders.

Organization of this Book

The three chapters that follow this introduction (Chapters 2, 3, and 4) provide information about the foundational areas of SUI design with a focus on filling in the knowledge gaps for students and practitioners with a background in human factors engineering and human-computer interface (HCI) design. Many human factors engineers and most HCI designers have significant experience in the design and evaluation of graphical user interfaces (GUIs) but do not have similar experience with speech technologies, human spoken language and communication, or self-service technologies.

Chapter 2 ("Speech Technologies") provides basic background in speech recognition, speech production, and speech biometrics. Chapter 3 ("Key Concepts in Human Language and Communication") addresses topics such as implicit linguistic knowledge, phonology, coarticulation, prosody, conversational discourse, conversational maxims, grammaticality, discourse markers, timing and turntaking, and social considerations in conversation—summarizing research in the areas I have found most useful in making the transition from standard human factors engineering to SUI design. Chapter 4 ("Self-Service Technologies") contains a survey of research in this important component of service science, including background in service science, call centers, technology acceptance and readiness, self-service technology adoption in general and for IVRs in particular, waiting for service, service recovery, and the consequences of forced use of self-service. Readers who are expert in one or more of these areas can skip the foundation chapters as appropriate.

The remaining chapters cover topics more directly related to the development and design of SUI IVRs, starting at a general level and getting ever more detailed. Chapter 5 ("The Importance of Speech User Interface Design") summarizes market and scientific research on user acceptance of speech IVR applications, addressing the question of why enterprises should consider the use of speech technologies in their IVRs. Chapter 6 ("Speech User Interface Development Methodology") goes over the phases required to develop SUI IVRs, from concept through deployment and tuning. Chapter 7 ("Getting Started: High-Level Decisions") gets into the fundamental high-level design decisions associated with speech IVR applications, including barge-in, selection of input and output methods, prompting styles, global commands, use of human agents, and the design of help. Chapter 8 ("Getting Specific: Low-Level Design Decisions") moves to more detailed topics such as the design of introductions, timing issues, designing prompts and menus, and confirming spoken input. Chapter 9 ("From 'Hello World' to 'The Planets': Prototyping SUI Designs with VoiceXML") provides an extended exercise for prototyping a speech application, covering many of the topics addressed in Chapters 7 and 8. The book ends with Chapter 10 ("Final Words"), which includes a list of resources for monitoring future developments in SUI research and design practices.

Summary

SUI design for interactive voice response systems rests on three major foundational bodies of knowledge: speech technologies, human factors, and service science. Without speech technologies, SUIs could not exist. SUI design should draw upon research from human factors and service science (and their related fields), when available, to guide design. In support of that goal, Chapters 2 through 4 provide information about the foundational bodies of knowledge, and Chapters 5 through 9 provide guidance on the development and design of SUIs, drawing upon the research literature to as great an extent as possible. Chapter 10 wraps up the book and provides a list of resources for monitoring future developments in SUI research and design practices.

References

Bitner, M. J., Ostrom, A. L., & Meuter, M. L. (2002). Implementing successful self-service technologies. *Academy of Management Executive, 16*(4), 96–108.

Bloom, J., Tranter, L., Hura, S., Krogh, P., McKienzie, J., Parks, M. C., & Tucker, D. (2008). The role of data in voice interaction design. From the 2008 Workshop on the Maturation of VUI.

Hura, S. L. (2009). Are you working hard to suck less? *Speech Technology, 14*(2), 7.

Lewis, J. R. (2011). Human factors engineering. To appear in the *Encyclopedia of Software Engineering*.

Meister, D. (1999). *The history of human factors and ergonomics*. Mahwah, NJ: Lawrence Erlbaum.

Pieraccini, R., & Hunter, P. (2010). Voice user interface design: From art and science to art with science. In W. Meisel (Ed.), *Speech in the User Interface: Lessons from Experience* (pp. 57–62). Victoria, Canada: TMA Associates.

Polkosky, M. D. (2008). Air cover in the trenches. *Speech Technology, 13*(3), speech-techmag.com/Articles/Column/Interact/Air-Cover-in-the-Trenches-41420.aspx

Spohrer, J., & Maglio, P. P. (2008). The emergence of service science: Toward systematic service innovations to accelerate co-creation of value. *Production and Operations Management, 17*(3), 238–246.

Suhm, B. (2008). IVR usability engineering using guidelines and analyses of end-to-end calls. In D. Gardner-Bonneau & H. E. Blanchard (Eds.), *Human factors and voice interactive systems* (2nd ed.) (pp. 1–41). New York, NY: Springer.

Wickens, C. D., Lee, J., Liu, Y. D., & Gordon-Becker, S. (2004). *An introduction to human factors engineering* (2nd ed.). Upper Saddle River, NJ: Prentice Hall.

2

Speech Technologies

The purpose of this chapter is to provide the basic information that a practitioner (designer or assessor) needs to work with speech user interface designs. Describing the fundamental analyses and programming (computational linguistics) that make speech recognition and synthetic speech possible is beyond the scope of this book, much less this chapter. For that level of information, refer to books such as Jelinek (1997) for speech recognition, Manning and Schütze (1999) for natural language processing, Taylor (2009) for synthetic speech (text-to-speech) production, Vacca (2007) for biometrics, and Jurafsky and Martin (2009) for computational methods.

This chapter also includes a brief introduction to writing and interpreting VoiceXML (scripting) and ABNF (grammar) code. Readers who have no interest in these topics can skip them. Those who have more of an interest in these topics should also read Chapter 9 and consult a book on VoiceXML programming, such as Sharma and Kunins (2002).

Speech Recognition

Figure 2.1 shows the waveform for a recording of the phrase "speech user interface design." This is the kind of information that a speech recognizer has to work with. Take a look at the figure—can you see where the word boundaries are—where one word ends and the next begins? That's hard to do without a fair amount of practice, and even after practice it isn't easy. Furthermore, a recognizer has the task of identifying this as "speech user interface design" whether the speech is slow or fast, male or female, young or old, native or nonnative speaker of English (Benzeghiba et al., 2007). As Lee (2003, p. 378) stated, "It is clear that the speech signal is one of the most complex signals that we need to deal with."

For an illustration of the word boundaries, see Figure 2.2. In particular, look at the boundary between the end of "user" and the beginning of "interface." This blending of word endings and beginnings is due to coarticulation—the tendency of speakers to run words together—a tendency that presents additional challenges to speech recognition.

To accomplish this feat, current speech recognizers use endpoint detection and two statistical models—an acoustic model and a language model. The purpose of endpoint detection algorithms is to identify the beginning

FIGURE 2.1
Waveform of "speech user interface design."

and end of speech and to identify the occurrence of a silence timeout (usually 5–7 seconds of silence without speech). Basically an acoustic model is a model of the sounds of the vowels and consonants of a language. A language model provides information about the words and arrangements of words that the system can recognize. There are two types of language models in current commercial use: finite state grammars (FSGs) and statistical language models (SLMs).

Speech user interface designers don't usually care much about the specifics of endpoint detection or acoustic models, which are more the purview of computational speech scientists, as long as the models are working properly. The distinction between FSGs and SLMs is, however, of great importance. Typically, designers use FSGs for more directed speech recognition (for example, to capture responses to a prompt such as "Please say Yes, No, or Repeat") and SLMs for more robust natural language understanding (for example, to recognize responses to an open-ended prompt such as "How may I help you?").

Finite State Grammars

A key characteristic of FSGs is that they fully specify what words and word orders the grammar can recognize. It is possible in some situations to use automatic methods to create FSGs, but in many cases designers craft them by hand. Table 2.1 shows an example of an FSG designed to capture responses

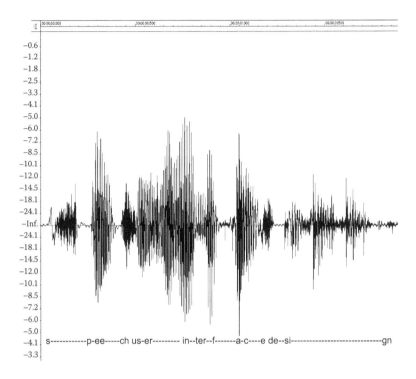

FIGURE 2.2
Waveform and text of "speech user interface design."

TABLE 2.1

Example of an FSG

```
#ABNF 1.0 iso-8859-1;
language en-us
mode voice;
root $account_type;

public $account_type = [$beginning]
        (checking {$="checking"} | savings {$="savings"})
        [$ending]
        [please];
$beginning = [I want] (my | the)
        | I want the balance of (my | the)
        | the balance of (my | the);
$ending = account
        | balance
        | account balance;
```

to the prompt, "Do you want the balance of your checking or savings account?" The example uses the augmented Backus-Naur form (ABNF), one of the grammar formats endorsed by the Worldwide Web Consortium (W3C, 2004a). At first its syntax might seem tricky, but with a little practice it's not that hard to read or write grammars suitable for use in prototypes (keeping in mind that development teams should include grammar experts who can produce more complex and highly tuned grammars). For example, let's go through the grammar in Table 2.1.

The first three lines identify the grammar format (ABNF), the language (U.S. English), and the mode (voice). The root line identifies the grammar name ($account_type). The next three sections (public $account_type, $beginning, and $ending) contain the rules that actually define the grammar, using the following symbols:

- To group items together, use parentheses—for example, "(my | the)"
- Vertical lines (|) mean "or"—for example, "(my | the)" means the user can say "my" or "the" at this place in the utterance, but not both
- Text in square brackets is optional—for example, "[please]"
- Curly brackets indicate semantic interpretation tags—for example, "{$="savings"}"

If a caller says anything that includes words or phrases that have semantic interpretation tags (which provide a concise and consistent interpretation of the meaning of the caller's utterance), the VoiceXML code has access to the utterance actually spoken by the caller and to any associated semantic interpretation tags. Thus semantic interpretation tags are important because they greatly simplify grammar output, which in turn simplifies the code written to interpret the output. For examples, see Table 2.2.

The public $account_type rule starts with an optional $beginning rule, followed by "checking" and "savings" options that include semantic interpretation tags, followed by an optional $ending rule and, optionally at the very end, the word "please." Table 2.2 shows various user responses to the prompt, "Do you want the balance of your checking or savings account?" along with

TABLE 2.2

Possible User Responses and Example Grammar Output

User Response	Recognized Text	Semantic Interpretation
checking	checking	checking
savings	savings	savings
my checking account, please	my checking account, please	checking
I want my savings account balance	I want my savings account balance	savings
How 'bout checking, OK	Invalid	No match

the output of the grammar. This grammar doesn't define everything that a user might say in response to the prompt, but it covers quite a bit of what users would probably say and is easy to modify to accept additional variations.

As shown in Table 2.2, if a user says a valid response (according to the grammar), then it will produce a semantic interpretation that is "checking" or "savings," even if the spoken phrase included more words (such as, "I want my savings account balance"). If the caller says something that is out of grammar (in this example, "How 'bout checking, OK?"), then the system will probably not recognize the input nor produce a useful semantic interpretation—or worse, might incorrectly match the input to portions of the grammar that have nothing to do with the checking account. To make this phrase in grammar, you could add "how 'bout" as an alternative in the $beginning rule and change "[please]" to "[please | OK]." The most common strategy is to make grammars only as flexible as they appear to require during initial development, then to add to or subtract from them as needed during iterative testing and tuning.

VoiceXML

Since its publication by the W3C, VoiceXML has become a very popular programming language for speech applications (Larson, 2003; Pieraccini & Lubensky, 2005; Sharma & Kunins, 2002; W3C, 2007). It is a markup language similar to HTML, the hypertext markup language used for many graphical sites on the World Wide Web, but with features to support speech user interfaces. Even though markup languages are programming languages, they operate at a high enough level that motivated nonprogrammers can use them to create prototypes of IVRs for the evaluation of initial designs. Table 2.3 shows a short working VoiceXML sample program that uses the grammar from Table 2.1 (account_type.gram), implementing the call flow shown in Figure 2.3.

Like HTML, VoiceXML is a computer language that uses tags, also called elements. VoiceXML doesn't have very many tags—fewer than 50 in Version 2.1. Because VoiceXML uses XML as its foundation, it requires a strict pairing of beginning and ending tags; for example, <tag> paired with </tag>, where the "/" before "tag" indicates "end." If it isn't necessary to put text between a beginning and ending tag, you can combine the beginning and ending tags by putting the "/" at the end of the tag before the closing angle bracket, for example, <tag/>.

The body of the program in Table 2.3 lies between the <vxml> and </vxml> tags, and includes four forms with the following IDs: "getaccounttype," "checking_info," "savings_info," and "goodbye." The first form (getaccounttype) starts by prompting the caller with, "Do you want the balance of your checking or savings account?" Below that prompt is a field named "account" that uses the grammar from Table 2.1 to recognize the caller's response to the prompt. The next section of the form lies between

TABLE 2.3

Example of VoiceXML

```
<?xml version="1.0" encoding ="iso-8859-1"?>
<!DOCTYPE vxml PUBLIC "-//W3C//DTD VOICEXML 2.1 WVS//EN" "vxml21-0728.dtd">

<vxml version="2.1" xmlns="http://www.w3.org/2001/vxml"
    xml:lang="en-US">
    <meta name="GENERATOR" content="IBM WebSphere Voice Toolkit" />

<form id="getaccounttype">
    <block>
        <prompt>Do you want your checking or savings balance? </prompt>
    </block>
    <field name="account">
        <grammar src="Grammars/account_type.gram"></grammar>
    </field>
    <filled>
        <if cond="account == 'checking'">
            <goto next="#checking_info"/>
        </if>
        <if cond="account == 'savings'">
            <goto next="#savings_info"/>
        </if>
    </filled>
</form>

<form id="checking_info">
    <block>
        <prompt>You've got ten thousand dollars in your checking account. </prompt>
        <goto next="#goodbye"/>
    </block>
</form>

<form id="savings_info">
    <block>
        <prompt>You've got one million dollars in your savings account.</prompt>
        <goto next="#goodbye"/>
    </block>
</form>
```

TABLE 2.3 (continued)

Example of VoiceXML

```
<form id="goodbye">
    <block>
        <prompt>Thanks for calling. Goodbye!</prompt>
        <exit/>
    </block>
</form>

</vxml>
```

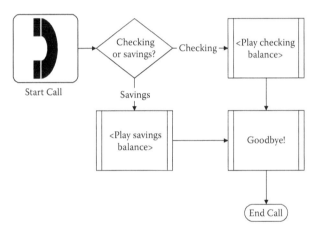

FIGURE 2.3
Graphic representation of program call flow in Table 2.3.

the <filled> and </filled> tags, and includes two <if> statements. If the output of the grammar is "checking," the call flow goes to the form with the ID "checking_info"; if it's "savings," the call flow goes to the "savings_info" form.

The "checking_info" form plays a static message about the checking account ("You've got ten thousand dollars in your checking account."), then directs the call flow to the "goodbye" form. The "savings_info" is identical to the "checking_info" form, except for the content of its static message ("You've got one million dollars in your savings account."). All the "goodbye" form does is to play, "Thanks for calling. Goodbye!" then the call ends (<exit/>).

Rather than writing a static VoiceXML program like the one in Table 2.3, professional programmers are more likely to use a programming language such as Java or C++ to dynamically generate VoiceXML code. Also a real program would need to get the balance information from a database and then integrate that information into the information played to the caller. On the other hand, even a VoiceXML program this limited could be used in early usability testing. See Chapter 9 for an extended exercise in prototyping with VoiceXML.

Statistical Language Models

Designing and creating statistical language models requires special skills and tools. This chapter does not cover the details of the necessary data gathering and manipulation. Rather, the focus is on high-level tasks and the operating characteristics of the technologies. Design teams using these technologies either need specialists on the team to perform these tasks or, if available, could purchase the models from vendors.

Like FSGs, SLMs fully specify what words they can recognize. Unlike FSGs, users can speak the words in any order. To provide a best guess at what a user said, recognizers use the SLM in combination with language modeling. To develop an SLM it is necessary to analyze a large amount of text that is representative of what users will say to the system. The resulting SLM uses the information from this analysis regarding the frequency of individual words (unigram frequencies), pairs of words (bigram frequencies), and, to some extent, triplets of words (trigram frequencies) (Jelinek, 1997). As discussed in more detail in Chapter 7, the flexibility of SLMs comes at a price of generally higher development and maintenance cost than FSGs (Balentine, 2010; Pieraccini, 2010).

To illustrate the operation of an SLM, suppose a caller said, "I'd like coffee with cream and … ." At this point, the SLM would have a strong bias to interpret the next word as "sugar." Strong evidence from the acoustic model could, however, overcome this bias. For example, if the user said, "that pink stuff," it is very unlikely that the recognizer would mistake this for "sugar."

Statistical Action Classification

When used in dictation programs, the result of the recognizer's output is typed text, with no additional interpretation. The previous section on FSGs discussed the use of semantic interpretation tags to simplify grammar output. The analogous method for SLMs is statistical action classification (SAC). To develop statistical action classifiers (SACs), developers reanalyze the text used to develop the SLM, this time assigning the various texts to categories of action.

After developing an initial SAC from a subset of the available text, developers can use that SAC to automatically classify the rest of the sentences and phrases, then check and revise the automatic classifications as required to get to the final SAC. Once the IVR is in use, a caller might say a phrase that was not in the original training data, but the SLM will attempt to recognize what the caller said and to classify it. Callers never see the text that the recognizer produces, so all the system needs is high enough recognition accuracy to enable correct action classification (Callejas & López-Cózar, 2008). Because these models have elements of chance there is always the possibility for

TABLE 2.4

Sample User Responses and Action Classifications

User Response	Action Classification
I'd like to test drive a hybrid.	Sales skill group
My engine's running rough.	Service scheduling IVR
My insurance company said to call you.	Collision skill group
I need to discuss something with you.	Unknown—disambiguation IVR

error. When properly developed, however, the combination of SLM and SAC can be very powerful.

One application of this combination is in natural language (NL) call routing (Kuo, Siohan, & Olive, 2003; Lee et al., 2000), a type of natural language understanding (NLU). Depending on what the caller says the NL call router directs the call flow to other self-service programs (speech, touchtone, or a combination) or to an appropriate call center skill group.

For example, imagine an NL call router for a car dealership. After calling the dealership's number, the IVR might start the conversation with a prompt such as, "Welcome to Cars 'R' Us. How may I help you?" Table 2.4 shows some possible responses to this prompt with associated action classifications.

Statistical Parsing

Statistical parsing takes NLU a step further. In addition to classifying the basic action or topic of the input speech, statistical parsing also attempts to extract meaningful units of information, called tokens. For example, consider the sentence, "I'd like to rent a van tomorrow at LAX." The desired action is to rent a car, but the sentence contains much more information than that. The type of car is "a van," the pickup date is "tomorrow," and the pickup location is "LAX."

As with other statistical language modeling, a statistical parser needs a lot of data and specialized skills and tools with which to build the model. Analysts go through the input texts and replace specific items with appropriate category names. After this analysis the example sentence above would look something like, "I'd like to rent <carType> <pickupDate> at <pickupLocation>" If the original sentence had been, "Can I pick up a Ford Focus from LAX tomorrow at 9 a.m.?" the new version would be, "Can I pick up <carType> from <pickupLocation> <pickupDate> at <pickupTime?>" The bigram and trigram language modeling uses the transformed rather than the original text because this produces a stronger model for both recognition and parsing. The content of the categories are kept in word lists or simple FSGs, which makes it easier to add new elements to the categories as appropriate. For example, if a company came out with new car models there would be no need to rebuild the statistical models—the developers would just add the new models to the <carType> list.

Dialog Management

In addition to the various statistical models they employ, NLU applications require dialog management (McTear, 2002; Wang, Yu, Ju, & Acero, 2008). The dialog manager (DM) for an NL call router must have conversational routines for dealing with speech that it cannot recognize, or statements that it can recognize but are inherently ambiguous. For example, suppose, in response to a request for a city name, a caller chewing gum said, "New York." If the recognizer wasn't sure about the caller's input, the DM might respond with something like, "Was that New York City or Newark, New Jersey?"

DMs for applications using statistical parsing must keep track of the information extracted so far and the information still required to complete the target transaction. Table 2.5 illustrates a caller's initial response and the resulting give-and-take between the caller and the DM for the car rental scenario. Even though in the initial utterance the caller provided all of the necessary information for picking up the car, the system needs to know the return information before being able to provide a quote.

Another characteristic of dialog management is the extent to which it allows a mixed-initiative conversation (Yang & Heeman, 2010). The interaction illustrated in Table 2.5 is single initiative because the IVR initiates all of the dialog turns by asking a question to which the caller responds. In a mixed-initiative mode, the caller can take the initiative by asking a question or making a statement. Suppose during the dialog shown in Table 2.5, when the IVR asks, "At what time?" the caller remembered having cancelled the credit card used in the caller's profile. In a mixed-initiative system, rather

TABLE 2.5

Getting the Information Needed to Quote a Car Rental

Speaker	Action
IVR	Majestic Car Rentals. How may I help you?
Caller	I'd like to pick up a van from LAX tomorrow at 9 a.m.
IVR	*<Extracts "van," "LAX," "tomorrow," "9 a.m.">*
	<Notes car type, pickup location, pickup date, and pickup time>
	<Computes pickup date for "tomorrow">
	<Begins collecting remaining required information>
	Dropping off at the same location?
Caller	Yes
IVR	Returning on what date?
Caller	The day after tomorrow
IVR	At what time?
Caller	Seven
IVR	AM or PM?
Caller	PM
IVR	OK, let's review …

than responding with the return time, the caller could say, "I need to change my credit card on file." After successful recognition of that request, an IVR supporting mixed initiative would switch to its credit card process, then, after completing that process, would return to finish capturing the remaining reservation information.

Confidence and N-Best Lists

At their core, all computational speech recognizers (FSGs and SLMs) use statistical modeling of one sort or another, with commercial systems predominantly using methods based on hidden Markov models (HMMs) (Baker et al., 2009; Bilmes, 2006). A speech recognizer, upon receiving a detectable audio input, will attempt to match it to candidates from its language model, whether that language model is an FSG or an SLM. At the same time it computes a measurement of how closely the input matches the candidate recognition results. This measurement of goodness-of-fit is the recognizer's confidence (Jiang, 2005; Torres, Hurtado, García, Sanchis, & Segarra, 2005). The list of recognition candidates, arranged in descending order of confidence, is the n-best list. If none of the members of the n-best list achieves a minimum level of confidence, the output of the recognizer indicates a nomatch condition. If at least one recognition candidate exceeds that minimum level of confidence, then the recognizer provides the item with the highest confidence as the recognized text. For a clever depiction of this process in the form of a series of cartoons, see Balentine (2007).

Detection of Emotion

Although no current commercial speech recognizers include emotion detection, researchers have explored the possibility of reliably detecting emotion from a caller's speech. Emotion detection has seen some use in mining call center recordings to find potentially emotionally charged interactions between callers and agents (Lee, Li, & Huang, 2008). Current estimates of the accuracy of emotion detection suggest, however, that it isn't ready for widespread use in IVRs (Lee & Narayanan, 2005; Ververidis & Kotropoulos, 2006). For example, Bosch (2003) reported difficulty distinguishing between anger and joy. With human accuracy of emotion detection from speech alone being about 60% (Scherer, 2003), it seems unlikely that automated detection could be much more accurate. Recently Pittermann, Pittermann, and Minker (2010) reported a two-step automated emotion detection accuracy of up to 78% for a test set in which human recognition was about 85%.

Furthermore, it isn't clear what value emotion detection would have in helping to design effective call flows beyond the elements that current recognizers can detect, such as the n-best list, recognition confidence measures, rejection, and silence timeout (Palen & Bødker, 2008; Rolandi, 2006), especially given its low accuracy (Batliner, Fischer, Huber, Spilker, & Nöth, 2003). For these reasons, emotion detection is not yet part of practical speech user interface design (Fenn et al., 2006).

Speech Production

There are two ways for applications to produce speech output. One is by playing recordings of human speech. The other is to produce synthetic speech with text-to-speech (TTS). TTS systems take text as input and produce speech as output with default pronunciations based on rules. If those rules mispronounce a word, then most systems allow for correcting the pronunciation in an exception dictionary. A major quality criterion for TTS systems is accuracy—the extent to which they correctly interpret the input text. Sometimes this requires knowledge of the context. After all, the correct pronunciation of "Dr." in "Dr. John Stevens" is very different from the "Dr." in the address "1016 Joy Dr." It is also important for the synthesizer to take into account punctuation to guide pausing and changes in fundamental frequency (for example, pitch falling at the end of statements and rising at the end of questions). The correct pronunciation rate of modern TTS synthesizers is generally high (for example, ranging from 98.9% to 99.6% in Evans, Draffan, James, & Blenkhorn, 2006).

To help developers annotate text to provide more accurate TTS, the W3C (2004b) has published a specification for a speech synthesis markup language (SSML). The SSML tags include <emphasis>, <break>, <prosody>, and <say-as>. The <emphasis> and <break> tags allow control over relative volume and the timing of pauses. The <prosody> tag affects the pitch and rate of speech. The <say-as> tag (available in some, but not all, VoiceXML development platforms) allows developers to specify the interpretation of the tagged text, for example, whether to interpret a number as currency or as a phone number.

Formant Text-to-Speech

There are two distinct types of TTS engines in current use: formant and concatenative. Formant TTS is the older technology, with generation of speech audio by rules based on the structure of the human vocal system (for example, the throat, tongue, and lips). The system requirements for formant TTS are relatively small, so it has been the method of choice for devices with memory and processing limitations. One of the pioneers of formant TTS, Dennis Klatt (1987), has made numerous samples of early TTS systems available on the Web (www.cs.indiana.edu/rhythmsp/ASA/Contents.html). Think, for example, of the sound of the synthetic voice associated with the physicist Stephen Hawking.

Concatenative Text-to-Speech

Concatenative TTS is a newer technology with generally better quality than formant TTS because it uses recordings of a human speaker to build a model for speech output (Aaron, Eide, & Pitrelli, 2005). The usual source of recordings is a predetermined set of phrases spoken by a professional voice talent. When a high-quality concatenative TTS system receives text input, it searches for the best match among its stored recordings. For this reason, larger databases of recordings lead to higher quality output.

For example, suppose the system needs to speak, "The weather in San Antonio will be chilly tomorrow." If this happened to be in the original set of recordings, the system could use that recording in its entirety. If the original recording had been, "The weather in Los Angeles will be warm this Tuesday," but the original recordings had included phrases such as "traveling to San Antonio tomorrow" and "chilly weather expected in January," then the system would extract the necessary recordings and would use computational rules to blend the edges of the recordings together as naturally as possible (in this example, using the segments "The weather in" "San Antonio" "will be" "chilly" "tomorrow").

For words that are not in the recording database, concatenative TTS systems generate the word from rules using diphone synthesis. This is an extreme form of the previously described blending, designed to mimic the coarticulation of sounds associated with natural human speech (see Chapter 3). The system builds the word by assembling diphones in which each diphone contains the second half of one speech sound and the first half of the next speech sound (Németh, Kiss, Zainkó, Olaszy, & Tóth, 2008).

Suppose the system needed to produce the word "will" between recordings of "The weather in San Antonio" and "be chilly tomorrow." The first diphone would need to blend as seamlessly as possible with the /o/ at the end of "Antonio" and would need to start the /w/ in "will," so it would be a diphone starting with the ending sound of /o/ and the beginning sound of /w/ given that the speaker has just produced /o/. The second diphone would start with the ending of a /w/ sound and the beginning of a short /i/ sound, and so on, until the system finishes producing the target word, taking into account the acoustic context in which it has appeared in the text.

Expressive Speech Production

Recent research in speech synthesis has explored the possibility of more expressive speech production (Pitrelli, Bakis, Eide, Fernandez, Hamza, & Picheny, 2006). For example, the speech signal produced when a speaker is smiling is different from that produced when a speaker is tense or angry. It might become possible to annotate text with tags that indicate different expressions. For example, "<smiling> That was fun. </smiling>" would produce the same words but with a different tone from "<ironic> That was fun.

TABLE 2.6

Components of Satisfaction with Speech Output

Component	Composed of Items Related to
Intelligibility	Listening effort, comprehension, articulation, pronunciation
Naturalness	Pleasantness, voice naturalness, human-like voice, voice quality
Prosody	Emphasis, rhythm, intonation
Social Impression	Trust, confidence, enthusiasm, persuasiveness

</ironic>". Once developed, these tags could be added to the W3C's SSML specification. For now, however, expressive synthetic speech production is not part of practical speech user interface design.

The Components of Satisfaction with Speech Output

For many years, the biggest problem with synthetic voices was their intelligibility (Francis & Nusbaum, 1999). Most current TTS systems, despite being more demanding on listeners than natural speech, are quite intelligible (Paris, Thomas, Gilson, & Kincaid, 2000). With the introduction and increasing sophistication of concatenative TTS, perceptions of naturalness, as measured with the standard Mean Opinion Score (MOS) questionnaire (Lewis, 2001; Mahdi, 2007), has become less of a discriminating factor. After developing an expanded version of the MOS that goes beyond the established factors of Intelligibility and Naturalness (Viswanathan & Viswanathan, 2005) to include items that measure prosodic and social–emotional aspects of speech, Polkosky and Lewis (2003) described four key components of satisfaction with speech output, as shown in Table 2.6.

Recorded Speech Segments

Surprisingly, it has turned out that the computational problem of producing convincingly natural and appealing synthetic voices is much more difficult to solve than the problem of getting reasonably accurate speech recognition. Computers don't care about the speech signals they receive when performing speech recognition, but people are very sensitive to the quality of speech that reaches their ears (Bailly, 2003). Some synthetic speech production errors are bizarre or jarring, while others convey paralinguistic signals that listeners reflexively interpret as lacking confidence or being untrustworthy. Executives of professional enterprises are reluctant to risk their brand image given the potential slips and other quality problems associated with even the best current synthetic speech.

For these reasons, most practical speech user interface designs use recorded speech for their output audio to as great an extent as possible. For high-quality output, designers minimize the number of splices (connections) between recorded audio segments and the potential for coarticulation problems at those splice points, and hire an agency to do the recordings with professional voice talents in a professional recording studio. Ideally someone who knows the details of the application design is present during the recording session(s) to coach the voice talent regarding the context and appropriate tone for the recordings. Most agencies have the capability for coaches to phone into the session.

Speech Biometrics

Biometrics in general refers to various techniques for identifying individuals based on physical (static) or behavioral (dynamic) traits (Vacca, 2007), typically for the purpose of providing an alternative to or supplementing passwords for access to secure information. Physical traits include fingerprints, facial geometry, irises, and body scent. Handwritten signatures and typing rhythm are examples of behavioral traits. The sound of a person's voice is due to a combination of physical (shape of vocal tract) and behavioral (voice pitch and speaking style) traits, all of which affect the data contained in a person's voiceprint.

There are a variety of methods for creating voiceprints for use in speech biometrics. Some systems require speakers to enroll using the exact phrase(s) that they will speak to the system. Other approaches allow speakers to enroll using a set enrollment script (or, in some cases, any text) to create general voiceprints that systems can use to identify or verify speakers.

For the process of speaker identification, the system has a set of voiceprints and uses them to attempt to identify who is speaking. For speaker verification, the speaker has claimed to be a particular person, and the system compares the speaker's voice with the stored voiceprint to attempt to determine if the speaker is who he or she claims to be. It is possible to combine speaker verification with other means of verification, such as asking the caller to answer personal identification questions with answers known to the system (for example, "What is your account number?"), then using speech recognition to assess the correctness of the response and voiceprint analysis for speaker verification (Kaushansky, 2006; Markowitz, 2010).

In general, biometric methods based on behavioral traits are less accurate than those based on physical traits. According to Jain and Pankanti (2008), speaker verification, which has both physical and behavioral components, is much more error prone than the purely physical methods based on a user's fingerprint, iris, or face. Toledano, Pozo, Trapote, and Goméz (2006) found speech verification to be more error prone than fingerprint scanning, but less error prone than signature verification. The use of speech biometrics in IVRs is appealing because phones have microphones built in—no need for other specialized input devices. Until speaker verification becomes much more accurate, however, it will be useful only for low-security applications unless combined with other verification methods.

Summary

In this chapter, we've covered the basics of three speech technologies used in IVR applications: speech recognition, speech production, and speech biometrics.

Figure 2.4 shows an overview of speech recognition processes. It begins with speech captured by the system. After the system detects the end of speech (endpoint detection), it proceeds with acoustic and language modeling. Language models can be relatively static (FSG) or statistical (SLM, possibly with action classification and statistical parsing for NLU applications). At this point the system has recognized the input and can use that recognition to direct the call flow. In more complex applications a statistical dialog manager directs the flow.

The quality with which a system speaks to a user is as important as the quality with which it listens. For most practical speech user interface applications, designers work with recorded speech, using synthetic speech only as necessary.

The most recent speech technology to see commercial application is speech biometrics. Although its accuracy is lower than that of physical biometrics such as fingerprint scanning, it has significant appeal in IVR design for speaker verification because it requires no additional hardware for the capture of the biometric. Figure 2.5 illustrates the process of speaker verification. The system captures a sample of the caller's speech and compares it to

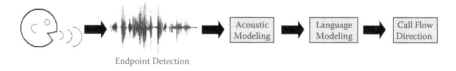

Endpoint Detection

FIGURE 2.4
Overview of speech recognition for IVR applications.

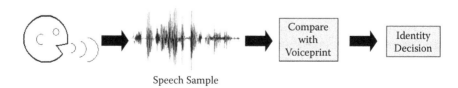

Speech Sample

FIGURE 2.5
Overview of speech verification for IVR applications.

the voiceprint of the person the caller is claiming to be, then uses the result of that comparison to decide whether the caller's identity claim is true.

References

Aaron, A., Eide, E., & Pitrelli, J. F. (2005). Conversational computers. *Scientific American, 292*(6), 64–69.

Bailly, G. (2003). Close shadowing natural versus synthetic speech. *International Journal of Speech Technology, 6*, 11–19.

Baker, J. M., Deng, L., Glass, J., Khudanpur, S., Lee, C., Morgan, N., & O'Shaughnessy, D. (2009). Research developments and directions in speech recognition and understanding, part 1. *IEEE Signal Processing Magazine, 26*(3), 75–80.

Balentine, B. (2007). *It's better to be a good machine than a bad person.* Annapolis, MD: ICMI Press.

Balentine, B. (2010). Next-generation IVR avoids first-generation user interface mistakes. In W. Meisel (Ed.), *Speech in the user interface: Lessons from experience* (pp. 71–74). Victoria, Canada: TMA Associates.

Batliner, A., Fischer, K., Huber, R., Spilker, J., & Nöth, E. (2003). How to find trouble in communication. *Speech Communication, 40*, 117–143.

Benzeghiba, M., De Mori, R., Deroo, O., Dupont, S., Erbes, T., Jouvet, D., Fissore, L., Laface, P., Mertins, A., Ris, C., Rose, R., Tyagi, V., & Welekens, C. (2007). Automatic speech recognition and speech variability: A review. *Speech Communication, 49*, 763–786.

Bilmes, J. A. (2006). What HMMs can do. *IEICE Transactions on Information and Systems, E89-D*(3), 869–891.

Bosch, L. (2003). Emotions, speech and the ASR framework. *Speech Communication, 40*, 213–225.

Callejas, Z., & López-Cózar, R. (2008). Relations between de-facto criteria in the evaluation of a spoken dialogue system. *Speech Communication, 50*, 646–665.

Evans, D. G., Draffan, E. A., James, A., & Blenkhorn, P. (2006). Do text-to-speech synthesizers pronounce correctly? A preliminary study. In K. Miesenberger et al. (Eds.), *Proceedings of ICCHP* (pp. 855–862). Berlin, Germany: Springer-Verlag.

Fenn, J., Cramoysan, S., Elliot, B., Bell, T., Davies, J., Dulaney, K., et al. (2006). *Hype cycle for human-computer interaction, 2006.* Stamford, CT: Gartner, Inc.

Francis, A. L., & Nusbaum, H. C. (1999). Evaluating the quality of synthetic speech. In D. Gardner-Bonneau (Ed.), *Human factors and voice interactive systems* (pp. 63–97). Boston, MA: Kluwer Academic Publishers.

Jain, A. K., & Pankanti, S. (2008). Beyond fingerprinting. *Scientific American, 299*(3), 78–81.

Jelinek, F. (1997). *Statistical methods for speech recognition.* Cambridge, MA: MIT Press.

Jiang, H. (2005). Confidence measures for speech recognition: A survey. *Speech Communication, 45*, 455–470.

Jurafsky, D., & Martin, J. H. (2009). *Speech and language processing* (2nd ed.). Upper Saddle River, NJ: Pearson Prentice Hall.

Kaushansky, K. (2006). Voice authentication—not just another speech application. In W. Meisel (Ed.), *VUI visions: Expert views on effective voice user interface design* (pp. 139–142). Victoria, Canada: TMA Associates.

Klatt, D. (1987). Review of text-to-speech conversion for English. *Journal of the Acoustical Society of America, 82,* 737–793. Audio samples available at www.cs.indiana.edu/rhythmsp/ASA/Contents.html

Kuo, H. J., Siohan, O., & Olive, J. P. (2003). Advances in natural language call routing. *Bell Labs Technical Journal, 7*(4), 155–170.

Larson, J. A. (2003). VoiceXML and the W3C speech interface framework. *IEEE Multimedia, 10*(4), 91–93.

Lee, C.-H. (2003). On automatic speech recognition at the dawn of the 21st century. *IEICE Transactions on Information and Systems, E86-D*(3), 377–396.

Lee, C.-H., Carpenter, B., Chou, W., Chu-Carroll, J., Reichl, W., Saad, A., & Zhou, Q. (2000). On natural language call routing. *Speech Communication, 31,* 309–320.

Lee, C. M., & Narayanan, S. S. (2005). Toward detecting emotions in spoken dialogs. *IEEE Transactions on Speech and Audio Processing, 13*(2), 293–303.

Lee, F., Li, L., & Huang, R. (2008). Recognizing low/high anger in speech for call centers. In *Proceedings of the 7th WSEAS International Conference on Signal Processing, Robotics, and Automation* (pp. 171–176). Stevens Point, WI: WSEAS.

Lewis, J. R. (2001). Psychometric properties of the Mean Opinion Scale. In *Proceedings of HCI International 2001: Usability Evaluation and Interface Design* (pp. 149–153). Mahwah, NJ: Lawrence Erlbaum.

Mahdi, A. E. (2007). Voice quality measurement in modern telecommunication networks. In *Systems, Signals and Image Processing, 2007,* and *6th EURASIP conference focused on speech and image processing, multimedia communications and services* (pp. 25–32). Maribor, Slovenia: IEEE.

Manning, C. D., & Schütze, H. (1999). *Foundations of statistical natural language processing.* Cambridge, MA: MIT Press.

Markowitz, J. (2010). VUI concepts for speaker verification. In W. Meisel (Ed.), *Speech in the user interface: Lessons from experience* (pp. 161–166). Victoria, Canada: TMA Associates.

McTear, M. F. (2002). Spoken dialogue technology: Enabling the conversational user interface. *ACM Computing Surveys, 34*(1), 90–169.

Németh, G., Kiss, G., Zainkó, C., Olaszy, G., & Tóth, B. (2008). Speech generation in mobile phones. In D. Gardner-Bonneau & H. E. Blanchard (Eds.), *Human factors and voice interactive systems* (2nd ed.) (pp. 163–191). New York, NY: Springer.

Palen, L., & Bødker, S. (2008). Don't get emotional. In C. Peter & R. Beale (Eds.), *Affect and emotion in HCI, LNCS 4868* (pp. 12–22). Heidelberg, Germany: Springer-Verlag.

Paris, C. R., Thomas, M. H., Gilson, R. D., & Kincaid, J. P. (2000). Linguistic cues and memory for synthetic and natural speech. *Human Factors, 42,* 421–431.

Pieraccini, R. (2010). Continuous automated speech tuning and the return of statistical grammars. In W. Meisel (Ed.), *Speech in the user interface: Lessons from experience* (pp. 255–259). Victoria, Canada: TMA Associates.

Pieraccini, R., & Lubensky, D. (2005). Spoken language communication with machines: The long and winding road from research to business. In M. Ali and F. Esposito (Eds.), *Proceedings of IEA/AIE 2005* (pp. 6–15). Heidelberg, Germany: Springer-Verlag.

Pitrelli, J. F., Bakis, R., Eide, E. M., Fernandez, R., Hamza, W., & Picheny, M. (2006). The IBM expressive text-to-speech synthesis system for American English. *IEEE Transactions on Audio, Speech, and Language Processing, 14*(4), 1099–1108.

Pittermann, J., Pittermann, A., & Minker, W. (2010). Emotion recognition and adaptation in spoken dialogue systems. *International Journal of Speech Technology, 13,* 49–60.

Polkosky, M. D., & Lewis, J. R. (2003). Expanding the MOS: Development and psychometric evaluation of the MOS-R and MOS-X. *International Journal of Speech Technology, 6,* 161–182.

Rolandi, W. (2006). Detecting emotion. *Speech Technology, 11*(3), 46.

Scherer, K. R. (2003). Vocal communication of emotion: A review of research paradigms. *Speech Communication, 40,* 227–256.

Sharma, C., & Kunins, J. (2002). *VoiceXML: Strategies and techniques for effective voice application development with VoiceXML 2.0.* New York, NY: John Wiley.

Taylor, P. (2009). *Text-to-speech synthesis.* Cambridge, UK: Cambridge University Press.

Toledano, D. T., Pozo, R. F., Trapote, Á. H., & Gómez, L. H. (2006). Usability evaluation of multi-modal biometric verification systems. *Interacting with Computers, 18,* 1101–1122.

Torres, F., Hurtado, L. F., García, F., Sanchis, E., & Segarra, E. (2005). Error handling in a stochastic dialog system through confidence measures. *Speech Communication, 45,* 211–229.

Vacca, J. R. (2007). *Biometric technologies and verification systems.* Burlington, MA: Elsevier.

Ververidis, D., & Kotropoulos, C. (2006). Emotional speech recognition: Resources, features, and methods. *Speech Communication, 48,* 1162–1181.

Viswanathan, M., & Viswanathan, M. (2005). Measuring speech quality for text-to-speech systems: Development and assessment of a modified mean opinion score (MOS) scale. *Computer Speech and Language, 19,* 55–83.

W3C. (2004a). Speech recognition grammar specification, version 1.0. Available at htpp://www.w3.org/TR/speech-grammar/

W3C. (2004b). Speech synthesis markup language (SSML), version 1.0. Available at htpp://www.w3.org/TR/speech-synthesis/

W3C. (2007). Voice extensible markup language (VoiceXML), version 2.1. Available at htpp://www.w3.org/TR/voicexml21/

Wang, Y., Yu, D., Ju, Y., & Acero, A. (2008). An introduction to voice search. *IEEE Signal Processing Magazine, 25*(3), 29–38.

Yang, F., & Heeman, P. A. (2010). Initiative conflicts in task-oriented dialogue. *Computer Speech and Language, 24,* 175–189.

3

Key Concepts in Human Language and Communication

As was the case for the preceding chapter on speech technologies, providing a comprehensive description of the various aspects of human language and communication—from the mechanics of speech production to speech comprehension to the social use of language—is beyond the scope of this book, much less this chapter. For a readable and entertaining introduction to linguistics and psycholinguistics, see Pinker (1994). Clark (1996, 2004) provides information on pragmatics, and West and Turner (2010) introduce communication theory. This chapter focuses on 10 concepts that are useful for practitioners who work with speech user interface (SUI) designs.

Because the language I speak and in which I primarily work is American English, this section has a strong focus on American English. All the languages of the world abound in subtleties of implicit linguistic knowledge and appropriate social nuances. SUI designers working on projects that include languages in which they are not fluent should ensure that they have access to fluent speakers to create and review the design (Elizondo & Crimmin, 2010; Klie, 2010).

Implicit Linguistic Knowledge

There are things that people know about the language they speak, but they don't know that they know. One way to illustrate this is with the "wug" test, shown in Figure 3.1. Unless you've studied psycholinguistics, you've probably never heard of a "wug" or a "bik." The wug test dates back to 1958 (Berko). It was used to study the implicit linguistic knowledge of children. It was possible that children, confronted with words they'd never before seen or heard, would refuse to speak them in pluralized form, but that's not what happened. In a significant majority of cases, the children pluralized "wug" and "bik" consistent with the rules of English word formation (morphology).

Try it yourself, and speak the words aloud. What did you do?

In both cases, if you were writing the answer in the blanks, you'd add an "s." When you said "biks," you most likely spoke an /s/. But when you spoke "wugs," you most likely actually spoke a /z/ rather than an /s/. To check, hold your fingers to your throat as you say "wugs" and "biks."

This is a wug.   Here is another one.

There are two _____ .

This is a bik.   Here is another one.

There are two _____ .

FIGURE 3.1
The wug test.

A voiced sound is one in which the vocal cords vibrate, creating a buzz that you can feel when touching your throat. The difference between a spoken /s/ and a spoken /z/ is that the /z/ is voiced and the /s/ is unvoiced. The difference between a spoken /k/ and a spoken /g/ is the same—the /k/ is unvoiced and the /g/ is voiced. This is not a coincidence. Speakers who follow the rules of English morphology pay implicit attention to the voicing of the preceding sound and use that to determine whether they speak an /s/ or a /z/ when pluralizing nouns. They follow the same rule when adding a /d/ or a /t/ to a verb to change it to past tense.

Speakers of Spanish do not follow this rule, for example, adding an /s/ to "gato" to make it the plural "gatos." A native speaker of English might mispronounce "gatos" by adding a /z/ instead of an /s/, whereas a native speaker of Spanish might mispronounce "figs" by adding an /s/ instead of a /z/. This is one way that native speakers can detect nonnative speech.

One implication of this rule for designers is that you must not mistake the properties of written language for those of spoken language. This is one reason why you can't record a single instance of /s/, then use it to pluralize both "dollar," which ends with a voiced consonant, and "cent," which ends with an unvoiced consonant. A more general implication is the value of SUI designers having a background in speech science to inform their design decisions.

Phonology

Phonology is the study of the basic sounds of language (phonemes) and the rules for their combination. The phonemes of a language include its vowels and consonants. When spoken, vowels tend to be longer than consonants, providing the foundation for the melody (prosody) of speech, with many consonants acting as percussive clicks and pops. This happens because vowels are produced with an open airway, but the production of consonants involves constricting the airway, either with the lips, tongue, or both (Fromkin, Rodman, & Hyams, 1998).

The articulatory phonetics of vowels depend on (1) the height of the tongue (high, middle, or low), (2) which part of the tongue is raised or lowered (front, central, or back), and (3) the position of the lips (rounded or unrounded). Languages differ in their inventory of vowels, averaging about 9 and ranging between 3 and 46 (Maddieson, 1984). In any given language, the distribution of vowel characteristics is such that the vowels tend to maximize acoustic distinctiveness (Stevens, 1989).

Languages tend to have more consonants than vowels, averaging around 23 and ranging from 6 to 95 (Maddieson, 1984). For consonants, the distinguishing features depend on (1) the place of articulation, (2) the manner of articulation, and (3) voicing. Places of articulation include bilabial (lips pressed together), labiodental (upper teeth against lower lip), interdental (tip of tongue between upper and lower teeth), alveolar (tongue touching the alveolar ridge, just behind the upper teeth), velar (back of tongue touching the soft palate), and palatal (front part of tongue touching the hard palate just behind the alveolar ridge). Manners of articulation include stops (complete blockage of the airway at some point) and continuants (partial blockage of the airway). Finally, consonants can be voiced or unvoiced. For example, /s/ and /z/ have the same place of articulation (alveolar) and manner of articulation (a type of continuant called fricative, made by blowing air through the obstruction), but, as described in the previous section, differ in voicing (/s/ is unvoiced, /z/ is voiced). /b/ and /d/ are both voiced stops, but differ in their place of articulation (/b/ is bilabial, /d/ is alveolar).

These distinctions can come into play for application designers when tuning speech recognition or artificial speech production. When a recognizer consistently fails to correctly recognize an in-grammar utterance, it is possible that the untuned phonemic analysis of one of the words does not correspond to the user's pronunciation. On the speech output side, it may be necessary to tune the pronunciation of a consistently mispronounced word for a particular application. Figure 3.2 shows a scheme (based on the IPA—International Phonetic Alphabet) for representing the vowels and consonants of English, with examples of the words in which they appear (20 vowels, 24 consonants).

Coarticulation

In natural, continuous speech, the articulators (tongue and lips) approach their articulatory targets but do not reach them. If speakers took the time

Vowels

Symbol	Example	Symbol	Example
ɑ	father	i	greed
ʌ	cup	ɪ	hit
æ	last	ɪə	spear
aɪ	hide	oʊ	hose
aʊ	ouch	ɒ	top
eɪ	late	ɔɪ	boy
e	present	u	aloof
ɜ	burn	ʊ	put
ʊə	sure	ə	away
eə	fair	ɔ	tall

Consonants

Symbol	Example	Symbol	Example
b	bed	r	rag
d	den	s	sip
f	file	ʃ	shell
g	guard	t	talk
h	hello	tʃ	chess
j	yes	θ	thin
k	kit	ð	there
l	lean	v	valley
m	milk	w	will
n	no	z	zoo
ŋ	long	ʒ	just
p	pull	dʒ	azure

FIGURE 3.2
IPA symbols for the vowels and consonants of American English.

to force the articulators to their final positions, their speech would sound artificial, and would be much slower than natural speech. The actual sound of phonemes depends on the phonemes that surround them, a phenomenon called "coarticulation." In some cases, the sounds of the phonemes blend in a way that makes it impossible to untangle them by simply viewing a picture of the resulting waveform (see Figure 2.2).

Despite this, our brains effortlessly untangle this complex waveform in real time. A related phenomenon is categorical perception—listeners perceive speech sounds that differ along a continuum as discrete categories of sounds (Liberman, Harris, Hoffman, & Griffith, 1957). In English, there are three pairs of stop consonants that differ only in the onset of voicing (with the first item in each pair voiced): /b/ and /p/ (bilabial), /d/ and /t/ (alveolar), and /k/ and /g/ (glottal) (Yeni-Komshian, 1993). For words that begin with voiced stops, the vocal cords vibrate before the tongue opens the airway for the following vowel. When experimenters artificially manipulate the amount of voicing before the vowel (voice onset time), there is no point at which listeners report being unable to identify the consonant, or even hearing the sounds as different "flavors" of the consonants. Around the crossover point, listeners report hearing the voiced consonant on the side with slightly greater voice-onset time, and report hearing the unvoiced consonant on the other side. At the crossover point, listeners sometimes report hearing the voiced consonant and sometimes report hearing the unvoiced consonant, but on any given instance confidently report what they heard.

An understanding of the concept of coarticulation is very important when planning the recorded output of a speech application. To as great an extent as possible, designers should plan for segments that are complete phrases to avoid having to deal with coarticulation. If this isn't possible, then coarticulation considerations should influence the recording plan to ensure that the recorded segments flow smoothly into one another.

Prosody

Prosody refers to the intonational pattern of a sentence or phrase—the pattern of pitches, stresses, and timing (Wingfield, 1993). These patterns can aid interpretation through emphasis of portions of the utterance and can provide cues to the emotion of the speaker. The words of a sentence may say one thing, but the prosodic delivery can reverse its interpreted meaning. Consider the sentence, "He's such a great humanitarian." Spoken as a normal sentence the words mean just what they say. Spoken sarcastically, the meaning is the opposite.

In English, the prosodic contour at the end of a sentence or phrase provides supplementary information to help listeners interpret the utterance as a statement or a question (Ratner & Gleason, 1993). At the end of a statement, the pitch (fundamental frequency) of the speaker's voice typically falls. At the end of a question, it usually rises. When spoken within a phrase, the pitch tends to stay the same.

Like coarticulation, prosodic effects can prevent two juxtaposed segments of recorded speech from sounding natural. For example, suppose you're planning to end some sentences with a dynamically computed date, sometimes finishing a statement and other times asking a question. It is important to account for these differences when putting together a recording plan.

Prosody also plays a role in the proper presentation of prompts and messages in a speech application. For any given sentence or phrase, there are many ways to speak it, only one or a few of which will be appropriate in a given situation. For example, what is the correct way to record the question (appearing in a list of frequently asked questions), "What happens after I apply for cash assistance?" Should the speaker emphasize "What," "happens," "after," "apply," or "cash assistance"?

The answer depends on the context in which the question appears. If the surrounding items concern other aspects of applying for and getting cash assistance, then you might want the speaker to emphasize "after," contrasting it with the things that happen before applying. If the surrounding items have to do with other types of assistance such as food stamps or health benefits, then you might want the speaker to emphasize "cash assistance." In short, it's important to get the prosodic element of contrastive stress correct (Cohen, Giangola, & Balogh, 2004).

Conversational Discourse

The field of linguistics, the scientific study of language, has enormous breadth. Linguists study language from the anatomical details of speech production in the individual (Hixon, Mead, & Goldman, 1976) to the changes that occur following the collision of cultures that speak different languages (Bickerton, 1981). Of particular value to the design of SUIs is pragmatics—the study of how people use language in everyday conversation to get things done (Clark, 1996). "Language evolved, after all, before people could read or write, attend plays, or watch television. Even today, the primary setting for language use is conversation" (Clark, 2004, pp. 365–366).

Most conversations follow a pattern. There must be an opening—someone must start the conversation. During the conversation, the speakers will take turns communicating with one another. There might be some overlap as one speaker ends and another begins, but in normal conversation there is relatively little overlap and relatively little silence at these transitional points. At the beginning, participants must establish their common ground (context) and if the topic changes must ensure that all participants are aware of the change. As the conversational participants speak, there are two systems of communication: primary and collateral. The primary system addresses the content necessary to complete the participants' joint activity. Since communication itself is an activity that requires coordination among participants, the collateral system includes the special communicative acts required for that coordination. Finally, the conversation ends with a closing. In addition to these activities, conversations have a number of requisites, as shown in Table 3.1.

The participants' focus of attention in a conversation is on the primary system of communication—on accomplishing the task at hand, whether that task is planning an activity or actually doing it. Participants are generally unaware of what they do to coordinate the conversation. Table 3.2 lists types of collateral signals used to coordinate conversation (Clark, 2004).

TABLE 3.1

Conversational Requisites

Requisite	Description
Participants	Those who will join in the conversation
Roles	The social roles of the participants—for example, peers, service provider, client, etc.
Content	The explicit goal of the conversation—for example, sharing stories about recent vacations, cashing a check, checking account balances, etc.
Timing	The time at which the participants will undertake the joint activities described in the conversation
Location	The place at which these activities will occur

TABLE 3.2

Collateral Signals

Signal	Description
Inserts	Bits of conversation that occur during the primary conversation, for example, to clarify the topic (side exchanges); short utterances ("yeah," "uh-huh") to encourage the current speaker to continue speaking (side moves); and fillers such as "uh" and "um" uttered by the current speaker to indicate that they intend to continue speaking, but need a little time to work out what to say next (asides)
Modifications	Speaking a word with a questioning intonation to indicate uncertainty (try marker); pronouncing "a" or "the" with a long vowel to indicate an upcoming pause in speech without the intention to give up the turn (nonreduced vowels); lengthening words in mid-utterance as the speaker plans the next part of the utterance (prolongation)
Juxtapositions	Repairing a word or phrase by saying it directly against the word or phrase it replaces (replacement); repeating a word just spoken (repetition); one speaker interrupts another in an attempt to claim the next turn (overlap)
Concomitants	Gestures that accompany speech, including gestures that depict the item or action to which they refer (iconic gestures) and those that indicate a desire to hand the turn to the other participant (turn gestures)

Current SUI design for IVR applications has a strong focus on providing service to callers. The wording of prompts and messages should be appropriate to those social roles, with careful attention paid to openings and closings. Designers need to conduct thorough task analysis to achieve the content goals of the interaction. When applicable, the system must be clear about any timing or location requirements for the successful completion of the task.

It's very difficult to plan for utterances in the collateral system when designing grammars. Fortunately, the operating characteristics of grammars allow a reasonable degree of flexibility in matching what callers actually say against what the grammar can recognize. For example, if a grammar can recognize the utterance "I'd like to rent a car," and the caller says "I, uh, I'd like to rent a car," then there's a good chance that the grammar will make the match. Also, for utterances such as "uh" and "um," many speech recognizers have "mumble" models that help the recognizer identify and ignore these types of fillers.

Conversational Maxims

In 1967, Paul Grice gave Harvard University's prestigious William James Lecture on the topic of logic and conversation (Grice, 1975). His primary goal was to explain the implications that people draw when they engage in conversation. In his introductory example, A and B are discussing a mutual friend C, who has taken a new job at a bank. A asks B how C is doing, and B says, "Oh quite well, I think; he likes his colleagues, and he hasn't been to prison yet" (p. 43).

There is a clear difference between the words that B has spoken and what those words imply. Grice noted that there is an assumption of a principle of cooperation among the participants in a conversation—in Grice's words (1975, p. 45), "Make your conversational contribution such as is required, at the stage at which it occurs, by the accepted purpose or direction of the talk exchange in which you are engaged." In the example above, there must be some reason that B has said, "and he hasn't been to prison yet." It might imply something about B's beliefs about bankers in general or about the likelihood that C will do something dishonest, but the fact that B spoke it indicates that it must have some meaning beyond the obvious. Participants in conversations do not add random facts to their statements. If B had said, "Oh quite well, I think; he likes his colleagues, and he didn't eat his socks for breakfast today," then A would spend some time trying to figure out what that was supposed to mean.

To provide a framework against which to identify utterances that require the interpretation of an implication, Grice (1975) presented his conversational maxims, summarized in Table 3.3.

If you're going to design a conversational application, then these maxims provide a reasonable (though incomplete) set of guidelines for straightforward communication. Failure to adhere to the maxims when no implication is intended opens the door to miscommunication with and mistrust of the application. In the example above, if A is unable to work out a reasonable interpretation of "and he didn't eat his socks for breakfast today," he might reasonably conclude that B is insane, and, in general, that you shouldn't trust people who are insane.

Moving beyond the obvious use of Grice's maxims as SUI design guidelines and building on a scale described by Bloom et al. (1999), Melanie Polkosky (2002) developed a pragmatic rating scale for the assessment of user-system dialogs. She had 96 participants listen to recorded dialogs from three different applications, with each application presented in three different styles:

TABLE 3.3

Conversational Maxims

Maxim	Description
Quantity	1. Be as informative as is required
	2. Do not be more informative than required
Quality	1. Do not make statements that you believe are false
	2. Do not make statements for which you lack evidence
Relation	Be relevant
Manner	1. Avoid obscure expressions
	2. Avoid ambiguity
	3. Be brief
	4. Be orderly

original (recorded human speech for prompts and messages), long TTS (same prompts and messages as in the original, but produced with a commercially available TTS engine), and short TTS (prompts and messages produced by TTS, but rewritten to be more concise). After psychometric analysis, the final rating scale had 11 seven-point items that aligned on four factors reasonably associated with the maxims.

She found higher ratings on the Quality scale for human prompts relative to the same prompts produced with TTS, and higher ratings on the Manner scale for short TTS compared to long TTS. Thus Polkosky's pragmatic scale (Table 3.4) discriminated between different aspects of SUI design in a manner consistent with Gricean theory. Future research in this area should investigate additional items to bolster the Quality and Quantity scales. Also, "while this

TABLE 3.4

Polkosky's Pragmatic Rating Scale for Dialogs

Factor	Items
Quantity	*Topic Changes*: Did the system respond appropriately when you asked about a different topic?
Quality	*Specificity*: Did the system provide specific, clear responses?
	Ambiguity: Were the system's responses ambiguous?
Relation	*Efficiency*: Did the system give information efficiently?
	Brevity: Were the system's responses brief?
	Speed: Did the system allow you to meet your goals quickly?
Manner	*Contingency*: Did the system provide responses based on your requests?
	Topic Development: Did the system's responses expand on your topic of conversation?
	Helpfulness: Was the system helpful?
	Predictability: Did your interaction with the system proceed as you expected?
	Consistency: Was the interaction with the system similar to other systems you have used (e.g., touch tone dial systems, face-to-face interactions, Web-based systems)?

method [having participants rate a dialog that they hear, but in which they are not participants] is consistent with the social psychological literature … future work should involve users of speech interfaces who interact directly with the application" (Polkosky, 2002, p. 11).

Grammaticality

When people speak to one another, do they speak grammatically? It depends on what you mean by "grammatical." Pinker (1994, p. 385) writes about the language mavens—"an informal network of copy-editors, dictionary usage panelists, style manual and handbook writers, English teachers, essayists, columnists, and pundits," who teach and write about prescriptive rules of English such as avoiding split infinitives or never ending a sentence with a preposition. What the layperson often thinks of as the "rules of English" started in the eighteenth century, with social forces that promoted modeling "proper" English grammar on the grammar of Latin, considered at that time the language of enlightenment and learning, and a way to indicate differences among social classes (Fromkin, Rodman, & Hyams, 1998).

Pinker (1994, p. 384) argues that the structural difference between Latin and English makes these types of rules fundamentally ridiculous or, at best, trivial. "When a scientist considers all the high-tech mental machinery needed to arrange words into ordinary sentences, prescriptive rules are, at best, inconsequential little decorations. The very fact that they have to be drilled shows that they are alien to the natural workings of the language system. One can choose to obsess over prescriptive rules, but they have no more to do with human language than the criteria for judging cats at a cat show have to do with mammalian biology." People engage in conversation effortlessly, but require more effort to produce or attend to speeches and other forms of monologs (Garrod & Pickering, 2004; Pickering & Garrod, 2004).

Much of the scientific work in linguistics relies on grammaticality judgments (Tremblay, 2005). Sometimes the judgments are those of participants in experiments (for example, Newport, 1991), and sometimes they are the intuitive judgments of linguists (Garrett, 1990). For example, consider the following sentences:

1. I put the car.
2. The horse raced past the barn fell.

The first sentence is ungrammatical because the verb "put" requires a location—"I put the car in the garage." The second sentence is a well-known garden-path sentence—a sentence that begins in a way that misleads readers—takes them down the garden path. When parsed correctly, it is grammatical. The easiest way to demonstrate the correct interpretation is with the analogous sentence, "The car driven past the barn stopped."

All of this type of linguistic research is fascinating in its own right, but it has little to do with how people talk to each other. To what extent are the utterances of people engaged in dialog grammatical, either in the prescriptive sense (following the "rules" of the language mavens) or in the descriptive sense (speaking sentences judged as grammatical either by naïve observers or professional linguists)? Consider the following exchange in which none of the utterances are grammatical sentences, yet the conversation is coherent.

A: How 'bout a bottle of Bud?

B: You 21?

A: No.

B: No.

Compare it with this exchange, in which all utterances are grammatical sentences.

A: May I have a bottle of Bud?

B: Are you 21?

A: No, I am not.

B: No, you may not have a beer.

Despite its lack of grammaticality (at least, at the sentence level), the first conversation is more natural than the second—more representative of a real-life conversation, and more consistent with the Gricean maxims. In the first conversation, the participants are as informative as required (but not more so), relevant, and brief—in other words, efficient. Interactions between call center agents and the callers they serve also tend to be efficient, as illustrated in Table 3.5 with a transcript from a recorded interaction.

Few of these utterances are grammatical sentences, especially after the caller and agent begin the give-and-take of the questions and answers needed to complete the reservation. They certainly could have spoken more words, but to do so would have adversely affected the efficiency of the communication. Expanding her earlier work on the Pragmatic Rating Scale for Dialogues, Polkosky (2005a, 2005b, 2008) developed a framework for SUI Service Quality with four factors (correlations with customer satisfaction shown in parentheses): User Goal Orientation (0.71), Speech Characteristics (0.43), Customer Service Behavior (0.40), and Verbosity (−0.27). All factors significantly correlated with customer satisfaction, with people preferring higher levels of the first three factors and lower levels of Verbosity.

Part of the guidance this research provides to the SUI designer is to embrace concise expression when scripting the dialogs for IVRs, within the

TABLE 3.5

Example of Call Center Interaction

Speaker	Action or Utterance
Caller	<Connects to call center>
Agent	*Good afternoon. Super Car Rentals. This is Ingrid.*
Caller	Hi, Ingrid. This is Susanna at Worldwide Travel. I need to book a car for a client.
Agent	*OK. So, for what city?*
Caller	Atlanta.
Agent	*And what day?*
Caller	April sixteenth.
Agent	*Incoming airline and flight?*
Caller	Airline is … American flight three oh six.
Agent	*That shows an arrival at eleven oh five?*
Caller	Correct.
Agent	*Returning when?*
Caller	April nineteenth.
Agent	*April nineteen, and what time?*
Caller	6 p.m.
Agent	*6 p.m., back to same location?*
Caller	Yes.

constraints of satisfying the other factors, especially User Goal Orientation. There is a tendency among novice designers to write dialog scripts that do not reflect the naturalness or efficiency of normal conversation because all the elements in their scripts are grammatical sentences. Avoid that tendency. When starting an IVR project, find the most skilled call center agents and listen to their interactions with callers (Yankelovich, 2008). Ideally, record and transcribe them. Use those interactions to guide (not to dictate, but to guide) the words and phrases used in the design.

Discourse Markers

One characteristic of conversation that aids efficient signaling about the intended direction of a conversation is the use of discourse markers—words whose main function is to indicate relations between the most recently spoken utterance and the upcoming utterance (Gleason & Ratner, 1993). Clark (1996) describes the use of discourse markers to indicate shifts in the topic of conversation. In Clark's terminology (p. 343), based on concep-tualizing conversations as joint projects, there are five types of transitions: entering the next project (Next), entering a subproject (Push), returning from the subproject (Pop), entering a digression from the main topic (Digress),

TABLE 3.6

Clark's Transition Types and Associated Discourse Markers

Transition	Examples of Associated Discourse Markers
Next	and, but, so, now, then, speaking of that, that reminds me, one more thing, before I forget
Push	now, like
Pop	anyway, but anyway, so, as I was saying
Digress	incidentally, by the way
Return	anyway, what were we saying

TABLE 3.7

Discourse Marker Types and Examples

Type	Examples of Associated Discourse Markers
Enumerative	first, second, next, then, finally
Reinforcing	also, furthermore, in addition, what's more
Equative	equally, likewise, similarly
Transitional	by the way, incidentally, now
Summative	then, in conclusion, to sum up
Apposition	namely, in other words, for example
Result	consequently, so, therefore, as a result
Inferential	else, otherwise, then, in other words, in that case
Reformulatory	better, in other words, rather
Replacive	alternatively, rather, on the other hand
Contrastive	instead, by comparison, on the other hand
Concessive	anyway, besides, however, nevertheless, still, after all
Temporal	meantime, meanwhile
Attitudinal	actually, strictly speaking, technically

and returning from that digression (Return). Table 3.6 contains examples of discourse markers associated with these types of conversational transitions. Cohen, Giangola, and Balogh (2004) provided a list of discourse markers, organized according to class of use, as shown in Table 3.7.

Speakers also use discourse markers to acknowledge what someone has just said, for example, "OK," "all right," and when appropriate, "thank you" or "thanks." Markers that signal a potential communicative problem are "oh," "hmm," and when appropriate, "sorry." These are among the most frequently used discourse markers in SUI design and, when properly used (not overused), enhance the natural flow of the prompts. Refer back to Table 3.5 to see the use of "OK," "so," and "and" as discourse markers in that natural dialog.

When using discourse markers in an IVR application, it's important to get the right intonation. For example, there are many ways to say "OK" or "oh," and these different ways can radically change the meaning of the utterance. "OK" spoken quickly and with neutral prosody indicates acknowledgment and acceptance. Drawn out and with a drop in pitch on the second syllable, it signals acknowledgment but suggests disbelief (Roberts, Francis, & Morgan, 2006). It is important that the voice talent be aware of the intended use of the discourse marker in context. If recorded separately, it is sometimes necessary to have multiple recordings of these types of discourse markers to ensure having the right one to use in a given situation.

Timing and Turntaking

"The right word may be effective, but no word was ever as effective as a rightly timed pause." (Mark Twain, quoted in Clark, 1996, p. 155)

Everything takes time. It takes time to produce speech, to understand speech, and to coordinate turntaking. There is no written script in normal conversation. We make it up as we go along, and timing is critical to the success of the interaction.

Typical speech production takes place at the rate of about 150–250 ms per syllable (Crystal & House, 1990; Massaro, 1975). This, along with other psycholinguistic research such as reaction time in response to mispronounced syllables, indicates that there is a limit of about 250 ms on the duration of the earliest stage of speech processing.

In face-to-face conversation, participants are adept at timing their utterances as they take their turns in the dialog (Clark, 1996), with one person talking at a time and, when they occur at all, brief silences, typically no longer than a second. Conversants try to minimize pauses at the end of a turn, while at the same time trying to avoid overlapping speech (Sacks, 2004). When a participant in a conversation pauses longer than 1 second and the other participant does not take the turn, it is common for the first participant to interpret this as indicative of a problem (Roberts, Francis, & Morgan, 2006). "Of course, interaction does not always run smoothly, but simultaneous talking, long silences, and other disruptions are notable precisely because they stand out against the background of normally unproblematic speech exchange" (Wilson & Zimmerman, 1986).

Even over the telephone, when nonverbal turntaking cues such as mutual gaze are not present, participants rarely talk over one another, with average overlap durations of 250 ms. When there are gaps in a telephone conversation, they tend to last less than 500 ms (Ervin-Tripp, 1993). Although other prosodic cues can come into play, one of the strongest verbal cues that a speaker wants to "yield the floor" (hand the turn to the other participant) is for that speaker to simply stop talking (Johnstone, Berry, Nguyen, & Asper, 1994; Margulies, 2005).

Beattie and Barnard (1979) investigated the timing of turns in telephone calls for directory inquiry. They found that the management of turntaking with regard to the timing of switching speakers was similar to that found in face-to-face conversation. Operators and callers had similar turntaking behavior despite their differences in experience. The mean switching time was 426 ms, with a 95% confidence interval ranging from 264–588 ms. The estimated 95th percentile was 1010 ms, and the estimated 99th percentile was 1303 ms.

These findings from human–human communication can inform research into and the design of SUIs. Pauses in system speech of about a quarter of a second (250 ms) should not typically trigger a turntaking behavior in a caller. During an ongoing dialog, system pauses longer than 1300 ms are very likely to induce a caller to begin speaking. It is probable that other specific characteristics of human–computer spoken dialog will also influence turntaking behavior, such as prosodic cues and differing dialog situations—asking open-ended questions, asking yes/no questions, providing a menu of choices, and so on. Effective SUI design requires careful specification and tuning of pauses to ensure an efficient and pleasant user experience.

Social Considerations in Conversation

Conversation does not take place in a vacuum. There must be at least two participants, and as soon as you have two people talking to one another, you have a social situation. Suppose two people are working together to solve a puzzle. The way they speak to one another will differ if they are parent and child, siblings, close friends, distant acquaintances, co-workers, or worker and manager.

As mentioned previously in the discussion of grammaticality, prescriptive rules can serve the social purpose of class identification. For example, the word "ain't" is not part of "proper" English, even though all speakers of English understand it and are perfectly capable of using it. Regular use of "ain't" in place of "isn't" reflects social aspects of education and class. The various ways in which we address one another reflect our social relationships. Slang and jargon can establish who is in and who is out of different social groups (Fromkin, Rodman, & Hyams, 1998).

Another aspect of social consideration in conversation is the directness of a request. Consider the following:

- Pour me a cup of coffee.
- Please pour me a cup of coffee.
- Would you please pour me a cup of coffee?
- Is there any more coffee?

All are requests for a cup of coffee, but they differ in directness, and consequently in the politeness of the request. The appropriate form for the request depends partly on how much the requestor wants the coffee, and more on the social relationship between the participants. Ask rudely from a position of little social power and you risk direct refusal with accompanying loss of face (Clark, 1996), and the loss of coffee. Ask too indirectly and you risk misinterpretation of the request and loss of coffee, but avoid loss of face.

Getting this aspect of tone correct plays an important role in creating a satisfactory interaction between a customer and a service provider (Polkosky, 2006). Less direct (more deferential) requests from the service provider imply greater choice on the part of the customer, which increases the customer's satisfaction with the service (Yagil, 2001). Customers might not be able to articulate why they perceive or fail to perceive an appropriate level of respect from a service provider, but they have the very human capacity to detect an inappropriate tone and have a corresponding negative emotional reaction.

Common politeness markers include phrases such as "could you" and "would you mind" (Ervin-Tripp, 1993). Even though the syntactic form these expressions take is that of a yes/no question ("Would you please pour me a cup of coffee?"; "Is there any more coffee?"), the clear implication is that these are requests. Only someone who is refusing to cooperate in the conversation, possibly as an indication of anger or an attempt at humor, would respond with a simple "yes" or "no" rather than just smiling and pouring the coffee.

Another way to avoid directness is to use passive voice. For example, compare these two sentences:

- Please brew some coffee.
- Some coffee should be brewed.

In English, sentences in passive voice (1) start with the object of the verb rather than the subject and (2) have some form of the verb "be" (or in some cases, "get") before the main verb. If the subject of the verb appears at all, it does so in a "by" prepositional phrase after the verb. Compare:

- The dog chased the cat. (active voice)
- The cat was chased by the dog. (passive voice)

Generally, writers should avoid passive voice (Purdue University Online Writing Lab, 2007). Passive voice leads to wordier, weaker writing—passive sentences rewritten as active can be as much as 40% shorter. Passive voice is vague, especially when the subject of the sentence doesn't appear at all—so the writing sounds evasive, avoiding responsibility. Finally overuse of passive voice can cause readers to lose interest. Readers prefer documentation with reduced use of passive voice (Lewis, 2006).

TABLE 3.8

Appropriate Uses of Passive Voice

Usage	Description
Focus	To put the focus on the object of the sentence.
	"That's the car that was parked by John."
Continuity	To achieve a smooth connection between the end of one sentence and the beginning of the next, especially in dialog.
	A: "Did John park **that** car?"
	B: "No, **this** car was parked by John."
Scientific writing	To avoid the use of personal pronouns ("I," "we") in scientific or other formal writing. Note that this practice has been changing, especially in human factors and psychology.
	"The expected effect was not found."
Common construction	Although Mrs. Smith did the work, we would normally say:
	"John Smith was born on January 5, 1984."
Obscure responsibility	To avoid indicating the responsible party.
	"Some coffee should be brewed."
Appropriate tone for service provider	To politely inform a customer of something they must do.
	"Check-in must be completed 30 minutes before the flight."

Furthermore, there is scientific evidence that supports the commonly given advice against the use of passive voice. Research in psycholinguistics and human factors has consistently shown that it is harder for people to extract the meaning from a passive sentence relative to its active counterpart (Broadbent, 1977; Ferreira, 2003; Garrett, 1990; Miller, 1962)—possibly taking 25% longer to understand a sentence expressed in passive voice (Bailey, 1989). Listeners appear to use different brain pathways when processing active and passive sentences—patients with Broca's aphasia (due to a specific type of damage to the left hemisphere) can accurately interpret active sentences but cannot accurately interpret passive ones (Zurif, 1990; Berndt, Mitchum, Burton, & Haendiges, 2004).

Given all this, one might wonder why passive voice exists in language. As shown in Table 3.8, there are several legitimate uses of passive voice. There is a structural rationale for focus, continuity, and in traditional scientific writing, to emphasize objects of verbs rather than subjects. In contrast to a structural rationale, the rationale for obscuring responsibility is more social. In addition to providing a way to make indirect (polite) requests, passive voice allows speakers to bring up potentially touchy topics in a relatively polite way, without explicitly identifying the responsible party. Designers should avoid indiscriminate use of passive voice in SUIs but should not fear to use it when it is appropriate.

Designers of SUIs should be sensitive to the social aspects of their designs. In many cases, differences of opinion among professional SUI designers with regard to the appropriate directness of requests made by speech applications have to do with different emphases on efficiency versus social considerations. Polkosky's (2005a, 2005b, 2008) framework for SUI Service Quality shows that efficiency and sociality are both important elements of SUI design, so designers should consider them both, striving to achieve high efficiency and high social competence.

Summary

From the physics of acoustics to human biology to social interactions, the study of language can and should inform the design of SUIs. Almost everyone fluently speaks at least one language, but there are many aspects of language that involve implicit knowledge, making it a nontrivial undertaking to apply linguistic knowledge to design. For example, a naïve approach to artificial speech production would be to have a speaker record a dictionary's worth of words, and then to play those recordings as required. A basic understanding of phonology, coarticulation, and prosody would enable one to predict the failure of such an approach.

A background in phonology, coarticulation, and prosody also provides useful knowledge when working on the plan for the recording manifest for an IVR application. Without this knowledge, designers run the risk of recording more segments than necessary, or the even greater risk of recording too few. For example, there might be a need for multiple recordings of the audio used to produce years in dates, depending on whether the audio will appear in more than one context (for example, at the end of a statement and at the end of a question). Incorrect prosody and bizarre juxtapositions are distracting and confusing to listeners.

It is a fundamental human factors engineering strategy to take advantage of population stereotypes (Lewis, 1986). In the United States, for example, there is a strong user expectation that power switches go up for "on" and down for "off." For SUI design, the population stereotypes are the users' linguistic expectations (both explicit and implicit). Thus effective script writing requires an understanding of conversational discourse, conversational maxims, real-world grammaticality, appropriate use of discourse markers, and timing.

Finally the social setting has a strong influence on other aspects of language use. For IVRs the most common social setting is that of service receiver (the caller) and service provider (the IVR). For the past decade there has been an increasing amount of research in service science, especially in marketing research of self-service technologies. This is another area of research that can inform the design of SUIs for IVRs, and is the topic of the next chapter.

References

Bailey, R. W. (1989). *Human performance engineering: Using human factors/ergonomics to achieve computer system usability.* Englewood Cliffs, NJ: Prentice-Hall.

Beattie, G. W., & Barnard, P. J. (1979). The temporal structure of natural telephone conversations (directory enquiry calls). *Linguistics, 17,* 213–229.

Berko, J. (1958). The child's learning of English morphology. *Word, 14,* 150–177.

Berndt, R. S., Mitchum, C., Burton, M., & Haendiges, A. (2004). Comprehension of reversible sentences in aphasia: The effects of verb meaning. *Cognitive Neuropsychology, 21,* 229–245.

Bickerton, D. (1981). *Roots of language.* Ann Arbor, MI: Karoma Publishers.

Bloom, R., Pick, L., Borod, J., Rorie, K., Andelman, F., Obler, L., Sliwinski, M., Campbell, A., Tweedy, J., & Welkowitz, J. (1999). Psychometric aspects of verbal pragmatic ratings. *Brain and Language, 68,* 553–565.

Broadbent, D. E. (1977). Language and ergonomics. *Applied Ergonomics, 8,* 15–18.

Clark, H. H. (1996). *Using language.* Cambridge, UK: Cambridge University Press.

Clark, H. H. (2004). Pragmatics of language performance. In L. R. Horn & G. Ward (Eds.), *Handbook of pragmatics* (pp. 365–382). Oxford, UK: Blackwell.

Cohen, M. H., Giangola, J. P., & Balogh, J. (2004). *Voice user interface design.* Boston, MA: Addison-Wesley.

Crystal, T. H., & House, A. S. (1990). Articulation rate and the duration of syllables and stress groups in connected speech. *Journal of the Acoustical Society of America, 88,* 101–112.

Elizondo, J. L., & Crimmin, P. (2010). Spanish, English, or … Spanglish? In W. Meisel (Ed.), *Speech in the user interface: Lessons from experience* (pp. 152–156). Victoria, Canada: TMA Associates.

Ervin-Tripp, S. (1993). Conversational discourse. In J. B. Gleason & N. B. Ratner (Eds.), *Psycholinguistics* (pp. 238–270). Fort Worth, TX: Harcourt Brace Jovanovich.

Ferreira, F. (2003). The misinterpretation of noncanonical sentences. *Cognitive Psychology, 47,* 164–203.

Fromkin, V., Rodman, R., & Hyams, N. (1998). *An introduction to language* (6th ed.). Fort Worth, TX: Harcourt Brace Jovanovich.

Garrett, M. F. (1990). Sentence processing. In D. N. Osherson and H. Lasnik (Eds.), *Language: An invitation to cognitive science* (pp. 133–176). Cambridge, MA: MIT Press.

Garrod, S., & Pickering, M. J. (2004). Why is conversation so easy? *Trends in Cognitive Science, 8*(1), 8–11.

Gleason, J. B., & Ratner, N. B. (1993). *Psycholinguistics.* Fort Worth, TX: Harcourt Brace Jovanovich.

Grice, H. P. (1975). Logic and conversation. In P. Cole & J. L. Morgan (Eds.), *Syntax and semantics, volume 3: Speech acts* (pp. 41–58). New York, NY: Academic Press.

Hixon, T. J., Mead, J., & Goldman, M. D. (1976). Dynamics of the chest wall during speech production: Functions of the thorax, rib cage, diaphragm, and abdomen. *Journal of Speech and Hearing Research, 19*(2), 297–356.

Johnstone, A., Berry, U., Nguyen, T., & Asper, A. (1994). There was a long pause: Influencing turn-taking behaviour in human-human and human-computer spoken dialogues. *International Journal of Human-Computer Studies, 41,* 383–411.

Klie, L. (2010). When in Rome. *Speech Technology, 15*(3), 20–24.

Lewis, J. R. (1986). Power switches: Some user expectations and preferences. In *Proceedings of the Human Factors Society 30th annual meeting* (pp. 895–899). Dayton, OH: Human Factors Society.

Lewis, J. R. (2006). Effectiveness of various automated readability measures for the competitive evaluation of user documentation. In *Proceedings of the Human Factors and Ergonomics Society 50th annual meeting* (pp. 624–628). Santa Monica, CA: Human Factors and Ergonomics Society.

Liberman, A. M., Harris, K. S., Hoffman, H. S., & Griffith, B. C. (1957). The discrimination of speech sounds within and across phoneme boundaries. *Journal of Experimental Psychology, 54,* 358–368.

Maddieson, I. (1984). *Patterns of sounds.* Cambridge, UK: Cambridge University Press.

Margulies, E. (2005). Adventures in turn-taking: Notes on success and failure in turn cue coupling. In *AVIOS 2005 proceedings* (pp. 1–10). San Jose, CA: AVIOS.

Massaro, D. (1975). Preperceptual images, processing time, and perceptual units in speech perception. In D. Massaro (Ed.), *Understanding language: An information-processing analysis of speech perception, reading, and psycholinguistics* (pp. 125–150). New York, NY: Academic Press.

Miller, G. A. (1962). Some psychological studies of grammar. *American Psychologist, 17,* 748–762.

Newport, E. L. (1991). Contrasting concepts of the critical period for language. In S. Carey & R. Gelman (Eds.), *The epigenesis of mind* (pp. 111–130). Hillsdale, NJ: Lawrence Erlbaum.

Pickering, M. J., & Garrod, S. (2004). Toward a mechanistic psychology of dialogue. *Behavioral and Brain Sciences, 27,* 169–226.

Pinker, S. (1994). *The language instinct: How the mind creates language.* Cambridge, MA: MIT Press.

Polkosky, M. D. (2002). *Initial psychometric evaluation of the Pragmatic Rating Scale for Dialogues (Tech. Report 29.3634).* Boca Raton, FL: IBM.

Polkosky, M. D. (2005a). Toward a social-cognitive psychology of speech technology: Affective responses to speech-based e-service. Unpublished doctoral dissertation. University of South Florida.

Polkosky, M. D. (2005b). What is speech usability, anyway? *Speech Technology, 10*(9), 22–25.

Polkosky, M. D. (2006). Respect: It's not what you say, it's how you say it. *Speech Technology, 11*(5), 16–21.

Polkosky, M. D. (2008). Machines as mediators: The challenge of technology for interpersonal communication theory and research. In E. Konjin (Ed.), *Mediated interpersonal communication* (pp. 34–57). New York, NY: Routledge.

Purdue University Online Writing Lab. (2007). *Active and passive voice.* Retrieved from http://owl.english.purdue.edu/handouts/print/grammar/g_actpass.html.

Ratner, N. B., & Gleason, J. B. (1993). An introduction to psycholinguistics: What do language users know? In J. B. Gleason & N. B. Ratner (Eds.), *Psycholinguistics* (pp. 1–40). Fort Worth, TX: Harcourt Brace Jovanovich.

Roberts, F., Francis, A. L., & Morgan, M. (2006). The interaction of inter-turn silence with prosodic cues in listener perceptions of "trouble" in conversation. *Speech Communication, 48,* 1079–1093.

Sacks, H. (2004). An initial characterization of the organization of speaker turn-taking in conversation. In G. H. Lerner (Ed.), *Conversation analysis: Studies from the first generation* (pp. 35–42). Philadelphia, PA: John Benjamins Publishing.

Stevens, K. N. (1989). On the quantal nature of speech. *Journal of Phonetics, 17,* 3–45.

Tremblay, A. (2005). On the use of grammaticality judgments in linguistic theory: Theoretical and methodological perspectives. *Second Language Studies, 24,* 129–167.

West, R., & Turner, L. (2010). Introducing communication theory: *Analysis and application* (4th ed.). Columbus, OH: McGraw-Hill Higher Education.

Wilson, T. P., & Zimmerman, D. H. (1986). The structure of silence between turns in two-party conversation. *Discourse Processes, 9,* 375–390.

Wingfield, A. (1993). Sentence processing. In J. B. Gleason & N. B. Ratner (Eds.), *Psycholinguistics* (pp. 199–235). Fort Worth, TX: Harcourt Brace Jovanovich.

Yagil, D. (2001). Ingratiation and assertiveness in the service provider-customer dyad. *Journal of Service Research, 3*(4), 345–353.

Yankelovich, N. (2008). Using natural dialogs as the basis for speech interface design. In D. Gardner-Bonneau & H. E. Blanchard (Eds.), *Human factors and voice interactive systems* (2nd ed.) (pp. 255–290). New York, NY: Springer.

Yeni-Komshian, G. H. (1993). Speech perception. In J. B. Gleason & N. B. Ratner (Eds.), *Psycholinguistics* (pp. 89–131). Fort Worth, TX: Harcourt Brace Jovanovich.

Zurif, E. B. (1990). Language and the brain. In D. N. Osherson and H. Lasnik (Eds.), *Language: An invitation to cognitive science* (pp. 177–198). Cambridge, MA: MIT Press.

4

Self-Service Technologies

Speech technologies appear in many different types of products and serve many different needs. The vast majority of IVR applications exist to provide service to the people who call them, making them a type of self-service technology (SST). An IVR might simply route calls to skill groups in a call center (Armistead, Kiely, Hole, & Prescott, 2002) or might allow callers to perform self-service, but most IVRs do a combination of routing and self-service. In a sense, routing is a type of self-service because rather than talking to a general operator for call direction, the caller must make choices to connect to the appropriate skill group. Thus in contrast to complete self-service, skill-based routing ends with the caller connected to a human to complete the requested service.

Service Science

Service science (Lusch, Vargo, & O'Brien, 2007; Lusch, Vargo, & Wessels, 2008; Pitkänen, Virtanen, & Kemppinen, 2008; Spohrer & Maglio, 2008) is an emerging interdisciplinary area of study focused on systematic innovation in service. Service industries make up more than 75% of the U.S. economy, covering areas such as health care, education, utilities, financial services, and government services (Larson, 2008). One of the key concepts of service science is that payment for performance defines service, in contrast to payment for physical goods. Other attributes of service are that it is time perishable, created and used simultaneously, and includes a client who participates in the coproduction of value. Table 4.1 summarizes three types of performance that are of interest to service providers.

As work in a service system evolves, there is a tendency to shift focus from talent to technology. As Spohrer and Maglio (2008, pp. 243–244) explain:

> The trick lies in understanding or predicting when or how each of the transitions may be made. In this model, the choice to change work practices requires answering four key questions: (1) Should we (what is the value)? (2) Can we (do we have the technology)? (3) May we (do we have authority or governance)? (4) Will we (is this one of our priorities)? For example, consider the way call centers have evolved over the decades.

TABLE 4.1

Three Types of Service-Related Performance

Type	Description
Talent	High talent performance (trained chef)—best for clients with requirements for unique customization
Environment	Superior environment (average chefs with good cookbooks and well-equipped kitchen)—best for flexibility in filling service employment positions
Technology	Self-service technology (Web site for ordering dinner)—best for increasing efficiency for standardized tasks

Early technology call centers in the 1970s were often staffed with the actual developers and key technologists who had developed a technology. This is sometimes still the case when calling a young start-up company for technology support. However, as demand rises, it makes sense to provide average performers with a superior environment (e.g., computers with a Frequently Asked Questions tool). Later, as demand continues to rise and competition increases, it may be possible to outsource or delegate the call center component of the business to a service provider in India. Finally, as technology advances, websites and automated speech recognition systems can provide automated or self-service assistance to clients with questions.

Clearly, IVRs play a role in this final stage—the automation of service (self-service), where the balance of investment of time in deriving value from the service has largely shifted to the client who is seeking service (Rowley, 2006). Furthermore, given this shift in responsibility, it is important to design systems that enhance customer efficiency, which then leads to enhanced customer attraction and retention (Xue & Harker, 2002). Table 4.2 lists potential advantages and disadvantages of self-service for organizations and customers (Beatson, Lee, & Coote, 2007).

TABLE 4.2

Potential Advantages and Disadvantages of Self-Service

	Advantages	Disadvantages
Organizations	Reduced labor cost	Resource-intensive investment
	Increased productivity	Loss of interpersonal contact
	Improved competitiveness	More difficult service recovery
	Technological differentiation	Loss of up-selling opportunities
	Consistent service delivery	Staff perception of job loss
Customers	Lower service delivery cost	Intimidating technology
	Increased availability (hours)	Concern about service recovery
	Increased availability (location)	Loss of interpersonal contact
	Increased control over process	

Call Centers

Very few IVRs provide only self-service. Most also route callers to specific skill groups of agents who work at a call center (or, if agents also communicate with customers using technologies such as e-mail or online chat, a contact center). Thus a major influence on the design of any IVR that routes callers is the design of the call center's skill groups and its goals, which can be a "cost center, profit center, key source of revenue, key source of frustration, strategic weapon, strategic disadvantage, source of marketing research, [or] source of marketing paralysis" (Bergevin & Wyatt, 2008, p. 1).

Estimates of the number of customer service representatives (CSRs) vary, but are probably about 2% of the total workforce in the United States (Yellin, 2009) and in the UK (Armistead et al., 2002), with about a quarter of these people working in the finance and insurance industries. Other industries with large numbers of CSRs are telecommunications, information technology, and utilities.

In her book *Your Call is (not that) Important to Us*, Emily Yellin (2009) provided a history of the evolution of the call center. From their beginnings in the late 1800s, telephony services have had a record of poor customer satisfaction. By the early 1900s, many places had laws against swearing at telephone operators, and in at least one case the telephone company took back its telephone and refused service to a customer who had said, "If you can't get the party I want, then you may shut up your damn telephone" (Yellin, 2009, p. 25).

The first switchboard operators were teenage boys—the same pool of employees who had operated the telegraphs that existed before telephony (Yellin, 2009). In their previous role, however, they had virtually no personal contact with the users of the system. Their generally poor behavior in the new setting (such as cursing at customers, yelling to one another as they tried to connect calls, and wrestling between calls) led to significant customer complaints. This led the telephone company to hire young women in place of the teenage boys, establishing an expectation held to the current day in the United States of female telephone operators.

Next came an increasing focus on efficiency. Once the system included telephone numbers (before this, operators had to memorize subscribers' names), operators answered calls with the short phrase, "Number, please." In 1882 the average wait time for a call connection was 5 minutes; in 1897 it was 45 seconds; and in 1900 it was 6.2 seconds. Shortly after 1900 the development of rotary telephones and automated switching technologies began to replace local operators. Through the 1950s operators, almost all women, continued to connect long-distance calls. After the introduction of direct dialing of long-distance calls, the role of the operator further declined, with their jobs focused on the handling of special services such as collect calls and directory assistance.

In part to address the relatively high cost of collect calls, toll-free numbers first became available in the 1960s, then were widely used in the 1980s and 1990s. Toll-free numbers made it easier for companies to manage the expense of providing a way for customers to communicate with the company, either for direct sales or after-sales support. The need to efficiently manage toll-free calls along with computer technologies that provided quick access to customer records led directly to the development of call centers. It was also during this time that the term "customer service" came into general use.

A constant pressure on call centers is cost reduction. In the United States the approximate cost for a call handled by an agent is $7.50. Calls outsourced to another country cost about $2.35 per call. Automated calls average around 32 cents per call (Yellin, 2009). The dependent and sometimes conflicting goals of efficiency (cost management) and customer satisfaction (Beckett, 2004) drive significant flux in the modern call center, with management often shifting focus from year to year on cost versus service.

Balentine and Morgan (2001) provided a list of the attributes of effective call center agents. The major attributes are:

- Assume the caller is busy.
- Be efficient when communication is good.
- Be helpful when progress is slower.
- Be courteous and polite, rarely apologize, and never blame the caller.

Rafaeli, Ziklik, and Doucet (2008) conducted a study of the impact of call center employees' customer orientation behaviors on customers' perceptions of service quality. They noted that although scripting interactions with customers can reduce the effects of differences in employees' abilities and attitudes, rigid adherence to scripts can jeopardize rather than improve the customer experience. From an analysis of 166 interactions at a retail bank call center that did not involve rigid scripting, they identified five types of helpful customer orientation behaviors (see Table 4.3).

TABLE 4.3

Frequency of Occurrence of Helpful Customer
Orientation Behaviors

Customer Orientation Behavior	% of Total Calls
None	45
Offering explanations/justifications	32
Anticipating customer requests	26
Providing emotional support	19
Educating the customer	17
Offering personalized information	8

Any given call that included customer orientation behaviors often included more than one, making it difficult to evaluate the effect of a specific type of behavior. Overall calls that included customer orientation behaviors tended to receive higher service scores. Splitting the data by the average call duration for the industry (3 minutes) showed that this effect was not present in shorter calls but was strong in longer calls. Longer calls that included these helpful behaviors received high scores, but those that did not include helpful behaviors received poorer scores.

One specific customer orientation behavior that appeared to play a significant role in enhancing the customer experience was to offer personalized information. This was a behavior similar to anticipating the customer request, with the difference that the information provided to the customer was not directly related to the problem at hand, but rather was more of a suggestion about how to get better service in the future, based on specific information about the customer. In the example given by Rafaeli et al. (2008), a customer called to transfer funds. After completing that task, the employee said, "OK, uh, also by the way, since you're a select customer, are you aware that we have a toll-free select banking phone number that you can use?"

The characteristics of call centers and effective call center agents can provide some guidance to the design of IVRs. Customer interactions with call center personnel are more expensive than interactions with an IVR. In general, call center tasks that are amenable to scripting are prime candidates for self service. Even if it is not possible to completely automate a task, there can be significant savings by automating the initial steps; for example, caller identification and authentication. To as great an extent as possible, an IVR should mimic the characteristics of effective call center agents, with scripting that is initially efficient, helpful when the caller appears to need assistance, but not so terse as to appear to be rude. Interaction designers should be aware of helpful customer orientation behaviors such as offering personalized information (typically at the end of a call) or justifying in advance the need to collect personal information (for example, during caller identification and authentication).

Technology Acceptance and Readiness

According to the Technology Acceptance Model (TAM) of Davis (1989), the primary factors that affect a user's intention to use a technology are its perceived usefulness and perceived ease of use. A number of studies support the validity of the TAM and its satisfactory explanation of end-user system

TABLE 4.4

The Dimensions of Technology Readiness

Dimension	Description
Optimism	Positive view of benefits of technology—e.g., use not limited to regular business hours, belief in increased control and efficiency
Innovativeness	Tendency for early adoption of technology—e.g., figures out new technologies without help
Discomfort	Feeling of loss of control and being overwhelmed by technology—e.g., belief that system designs are unusable, lack of knowledge
Insecurity	Distrust of technology—e.g., will not do business online

usage (Wu, Chen, & Lin, 2007). In the TAM, perceived usefulness is the extent to which a person believes a technology will enhance job performance, and perceived ease of use is the extent to which a person believes that using the technology will be effortless.

In 2000, Parasuraman introduced the Technology Readiness Index (TRI), a psychometrically qualified scale designed to measure the readiness of people to use new technologies, including SSTs. The construct of technology readiness refers to "people's propensity to embrace and use new technologies for accomplishing goals in home life and at work" (Parasuraman, 2000, p. 308). Parasuraman identified four dimensions related to this construct, summarized in Table 4.4 (with examples from Tsikriktsis, 2004). Investigation of the four dimensions indicated acceptable reliability (coefficient alphas from 0.72 to 0.82) and validity. The TRI subscales appeared to predict respondents' technology-related behaviors.

Tsikriktsis (2004) used the TRI to classify customers in the UK, expecting to replicate five groups defined by Parasuraman and Colby (2001): Explorers (highly motivated and fearless), Pioneers (desire benefits but aware of dangers), Skeptics (somewhat more focused on dangers than benefits), Paranoids (aware of benefits but highly focused on dangers), and Laggards (unwilling to adopt unless forced). Analysis of the UK sample revealed four rather than five groups, with the characteristics of Explorers (27% of the sample), Pioneers (20%), Skeptics (32%), and Laggards (21%)—a partial replication of Parasuraman and Colby. Figure 4.1 shows the percentages for adoption by these four groups of five different technologies from Tsikriktsis.

Although not discussed by Tsikriktsis (2004), it is clear that the typical order of listing the customer groups did not correspond to their use of technologies. For all five technologies the order of adoption was highest by Explorers, then Skeptics, then Pioneers, and finally Laggards. The groups had different demographic profiles, with Explorers and Skeptics (the higher-adopting groups) tending to be male, younger, and with greater incomes.

In contrast to the relatively high usage of ATMs in Tsikriktsis (2004), which averaged 82.2% across customer groups, usage of the other SSTs was much

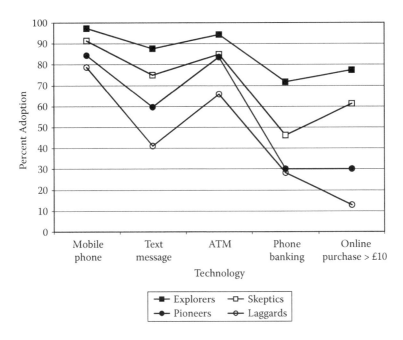

FIGURE 4.1
Percentage of technology adoption by TRI-defined customer groups.

lower. Usage of Web-based retail purchasing averaged 45.5% across customer groups, and adoption of phone banking averaged 44.1%, ranging from 28.2% for Laggards to 71.7% for Explorers—far from 100% for even the group with the highest adoption rate.

Recent research on the relationship between technology readiness and use of SSTs has produced mixed results. Using data from 413 customers of various SSTs, Lin and Hsieh (2007) confirmed the statistical reliability and validity of the four Technology Readiness (TR) subscales. The SSTs in their study included banks, railways, airlines, rapid transit systems, stock exchanges, and cinemas.

Liljander, Gillberg, Gummerus, and van Riel (2006) described an investigation of the relationship between technology readiness and the adoption of SSTs, specifically self-service airline check-in using kiosks and the Internet. They reported that adoption of kiosk usage had been low—despite its introduction at a major airport in 1996, by 2002 the adoption rate was just 14%. In constructing their questionnaire they drew upon the market research literature to address attitudes, adoption, quality, satisfaction, and loyalty. This questionnaire included 12 of the 36 original TR items (three for each of the four dimensions), selected on the basis of factor loading and relevance. Factor analysis of the TR data did not confirm the expected four dimensions—Optimism and Innovativeness emerged, but Discomfort and Insecurity were not reliably independent. Liljander et al. (2006) hypothesized that this failure to replicate could be due to the use of a subset of TR items, to the focus on a single task, or to cultural differences. Their key findings were:

- Only 58 (4.6%) of 1258 surveyed customers (80% of which were business travelers) had used Internet check-in; 473 (37.6%) had used kiosks.
- Their overall TR measurement and the Optimism subscale had consistent and significant association with customer attitudes toward use of these SSTs, but Innovativeness did not.
- None of the TR variables significantly predicted usage of the technologies.
- Those who reported wanting to use self-service check-in cited efficiency (time saved, avoid queues, ease and quickness) and being in control (especially control over when to check in) as primary reasons; the primary reasons given by those who avoided self-service were lack of perceived benefits, preference for personal service, inconvenience, and lack of trust (either in the automation or in one's ability to use the automation).

The research on TAM shows that users who adopt technologies perceive them as useful and easy to use. Studies of TR suggest that adoption might also depend on individual differences in Optimism, Innovativeness, Discomfort, and Insecurity, and that there may be discriminable clusters of users such as Explorers, Skeptics, Pioneers, and Laggards. But how does any of this research inform the design of an IVR?

One potential application of the research is to focus on IVR efficiency to avoid alienating those who are most likely to use the IVR. Attempts to increase IVR utilization by non-adopters often include lengthening prompts and providing more up-front promotion of the IVR's benefits—strategies that are not likely to satisfactorily address the reasons given by those who avoid self-service and which run counter to the desired efficiency of likely adopters (Liljander et al., 2006).

Another important finding is that even among customer groups who have relatively high TR, adoption of most SSTs falls far short of 100%. It is therefore unrealistic to expect 100% of customers to engage in 100% usage of SSTs, especially in extremely competitive markets where some companies provide (and advertise) personal service to attract those who avoid self-service. Alternatively markets served by a monopoly or near monopoly (or those in which the major market players have all invested in SSTs) can place greater emphasis on self-service. Companies that provide service to primarily male, educated, younger, and high-income customers can anticipate relatively higher adoption of their SSTs. An ongoing challenge to enterprises that provide service is to get the right balance between live service and self-service (Fluss, 2008).

Satisfaction with and Adoption of SSTs

The early 2000s saw a significant increase in market research of the drivers and inhibitors of customer satisfaction with self-service technologies. In a landmark paper, Meuter, Ostrom, Roundtree, and Bitner (2000) reported a critical incident study of more than 800 SST incidents gathered with an Internet survey. The technologies included IVR, Internet, and kiosks. Tasks included customer service (such as telephone banking, flight information, and order status), transactions (such as telephone banking and prescription refills), and self-help (such as information services). The respondents included about equal numbers of males and females, and were slightly younger, more educated, and with higher incomes than the general population—demographic characteristics associated with greater than average technology readiness (Tsikriktsis, 2004), not surprising given the Internet data collection method. Table 4.5 shows the resulting categories and percentages for satisfying incidents, and Table 4.6 shows the same for dissatisfying incidents.

Based on this research (Bitner, Ostrom, & Meuter, 2002; Ostrom, Bitner, & Meuter, 2002), Bitner et al. (2002) provided six key points for successful implementation of SSTs:

- Be very clear on the purpose of the SST: Is it primarily for cost reduction, customer satisfaction, competitive positioning, or some combination?
- Maintain a customer focus: Understand customer needs through interviews, surveys, and focus groups. Design for usability, and test to ensure a usable design.
- Actively promote the use of SSTs: Make customers aware of the SST. In a study of banking SST usage published by Curran, Meuter, and Surprenant (2003), 72.5% of banking customers had not used banking by phone—34.7% did not know if their bank offered the SST.
- Prevent and manage failures: Failure of technologies and service are the primary reasons why customers stop using SSTs. Because it is not possible to prevent all failures it is important to plan for service recovery.
- Offer choices: Customers expect to be able to interact with service providers using whatever method they prefer. Do not force usage of SSTs. Especially do not punish customers who try an SST by providing no option to communicate with a live person. If possible, provide incentives for SST use.
- Be prepared for constant updating and continuous improvement: An initial SST design will have room for improvement, perhaps due to improvements in technology, identification of additional opportunities for self-service, or changes in leading practice in the design of user interfaces.

TABLE 4.5

Drivers of Satisfaction with SSTs

Category	Description	% of Total
Better than alternative	The SST provided a benefit over traditional service, such as saving time (30%), easy to use (16%), when I want (8%), saving money (6%), where I want (5%), and avoiding service personnel (3%).	68
Did its job	Satisfaction emerged from an element of surprise that the technology worked as intended.	21
Solved intense need	Transactions that include a sense of urgency involve an intense need. The "always there" nature of SSTs enables immediate service when other service options are not available. This is especially powerful when coupled with an intense need.	11

TABLE 4.6

Drivers of Dissatisfaction with SSTs

Category	Description	% of Total
Technology failure	The technology simply didn't work, resulting in problems that would affect any user—e.g., broken ATM, Web site refusing to allow login.	43
Poor design	Problem with design that would affect some, but not necessarily all users—17% poor design of technology for usability (e.g., difficult enrollment or login procedure) and 19% poor service design (e.g., online purchase delivered to cardholder address rather than address of intended recipient of gift).	36
Process failure	Problem with process after successful completion of the initial customer–technology interaction—e.g., problem with billing or delivery.	17
Customer-driven failure	Problem in which customers believe they bear some responsibility for the failure—e.g., forgot PIN or password.	4

To better understand the attitudes affecting the intention to use SSTs, Curran et al. (2003) surveyed a random sample of homes with questions about attitudes toward bank employees and banking SSTs such as ATMs, phone banking, and Internet banking. About half of their sample used SSTs less than 25% of the time. They found two general attitudinal forces that affected the use of SSTs—the customer's attitude toward employees and the attitude toward SSTs. More experienced users of SSTs relied more on attitudes toward specific SSTs; less experienced users relied more on global attitudes toward SSTs.

Meuter, Ostrom, Bitner, and Roundtree (2003) investigated the relationship between technology anxiety (a construct conceptually related to the negative factors of the TRI) and usage of 14 types of SSTs. They found that technology anxiety was a better, more consistent predictor of SST use than

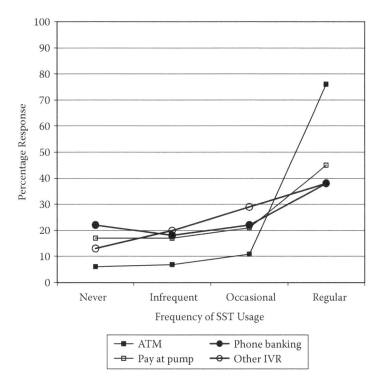

FIGURE 4.2
Patterns of common SST usage.

the demographic variables of age, gender, income and education. Figure 4.2 shows the pattern of usage of SSTs identified as being among the most frequently used. The patterns of phone banking and other IVR usage were similar, with 38% regular use and only 13% to 22% (other IVR phone banking) of respondents who never used them. In contrast, about 75% of respondents claimed regular use of ATMs.

In an independent study of banking SSTs, Curran and Meuter (2005) found similar patterns of usage, with 79.5% for ATMs and 27.5% for phone banking. In addition to the TAM factors of usefulness and ease of use, they also investigated the influence of a customer's need for interaction and perceived risk of the use of the technology on attitude toward the SST, which is a necessary precursor of the intention to use an SST. Structural equation modeling revealed that customers perceived ATMs as useful and easy to use, with neither the need for interaction nor risk affecting attitude. The pattern was similar for phone banking, except the only significant factor influencing attitude toward use was perceived usefulness. Thus one difference apparently affecting the different adoption rates of ATMs and phone banking is that of perceived usability—a factor known to affect behavioral intention of use (Strawderman & Koubek, 2008).

To further explore adoption of SSTs, Meuter, Bitner, Ostrom, and Brown (2005) collected survey data from customers who had recently placed prescription refill orders, some of whom used an IVR and others who used traditional mail order or spoke to a live agent. (At the time of the study, the company did not have a way to refill prescriptions over the Internet.) The results indicated that consumer readiness explained much of the variance in consumers' reluctance to try the SST. "Even customers who have a positive evaluation of an innovative service may choose not to use it if they do not understand their role (role clarity), if they perceive no clear benefit to using it (motivation), or if they believe that they are not able to use it (ability)" (p. 78). To increase the likelihood of getting customers to try a new SST, Meuter et al. (2005) recommended interventions based on these consumer readiness variables, such as wallet cards with detailed instructions (to address role clarity and ability) and clear communication of valued customer benefits (to address motivation).

Relationship of IVR to Other SSTs

Although they are all SSTs, ATMs, the Internet, and IVRs are very different from one another. They differ in the specific technologies used, their user interfaces, and the types of tasks that they can support. Cunningham, Young, and Gerlach (2008) used multidimensional scaling to study how customers perceived and classified a set of 12 SSTs—online banking, distance education, airline reservations, tax software, retail self-scanning, online auctions, pay at the pump, ATMs, online brokerage, IVR, Internet search, and car buying.

Participants rated each SST on the 11 features listed in Table 4.7. As shown in the table, analysis of the ratings led to an association of each feature with one of two dimensions. Figure 4.3 shows the mapping of SSTs on the multidimensional space defined by the dimensions of Separable/Inseparable and Customized/Standardized.

For Separable/Inseparable the key feature is Separability—the extent to which the production and consumption of the service occur at different times. For example, there is usually a distinct separation between booking a flight and taking the flight. The other features associated with the Separable/Inseparable dimension were Physical, Service Delivery, and Convenience.

For Customized/Standardized the primary feature was Customization—the extent to which the service process required significant customization versus more standardized service procedures. For example, the processes of paying at the pump or using an ATM are highly standardized. The other features associated with the Customized/Standardized dimension were Contact, Riskiness, Switching, Person/Object, Relationship, and Judgment.

TABLE 4.7

Service Classification Features

Feature	Description	Associated Dimension
Separability	Extent of separability of production and consumption of service	Separable/Inseparable
Physical	Extent to which SST is physical object	Separable/Inseparable
Service delivery	Extent to which service delivery is continuous or discrete	Separable/Inseparable
Convenience	Degree of convenience in obtaining service	Separable/Inseparable
Customization	Level of customization (versus standardization) of service	Customized/Standardized
Contact	Level of customer/employee contact	Customized/Standardized
Riskiness	Level of risk in choosing service provider	Customized/Standardized
Switching	Ease of switching to a new service provider	Customized/Standardized
Person/object	Extent to which service is performed on a person or tangible object	Customized/Standardized
Relationship	Level of formal relationship between customer and service provider	Customized/Standardized
Judgment	Extent to which service contact person can make service provision decisions	Customized/Standardized

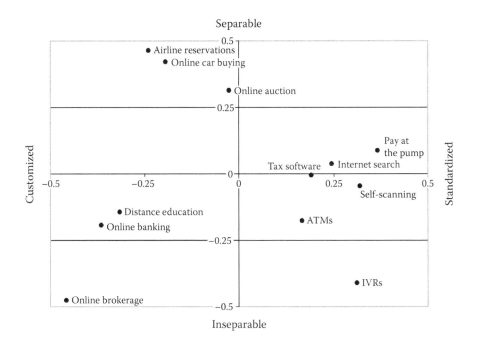

FIGURE 4.3

Customer perception and categorization of SSTs.

Although Figure 4.3 shows clustering of some SSTs, IVRs stand alone in the lower right portion of the graph, a location associated with a high degree of standardization and consumption of the service at the time of production (inseparability). These findings indicate that despite the common attributes of SSTs, their specific designs will need to match user expectations and abilities with the different technologies and associated tasks.

Waiting for Service

When customers want service they prefer immediate service, with some customers more sensitive than others to the perception of lost time (Kleijnen, de Ruyter, & Wetzels, 2007). Enterprises staffing for service must balance the cost of providing immediate service against the reduction in customer satisfaction due to waiting. Typically enterprises seek a balance that includes a tolerable wait time. What customers perceive as tolerable will differ as a function of the need for the service and the availability of competitors offering similar service.

In a theoretical investigation of the psychological cost of waiting, Osuna (1985) developed a mathematical model of the buildup of anxiety and stress due to the sense of waste and uncertainty inherent in a waiting situation. A key finding of the model was that the psychological cost of waiting is a marginal increasing function of waiting time, and just before people receive service the stress intensity is greater than it would be if they knew they were about to receive service. From this, it follows that there is value in providing information about expected wait times, and any delay in providing this information can cause unnecessary stress on the individual who is waiting.

When waiting, customers are sensitive to violations of social justice (Larson, 1987). In particular, customers expect adherence to FIFO (first in, first out). Because IVRs are usually single-queue rather than multiple-queue structures, they tend to enforce FIFO (Rafaeli, Barron, & Haber, 2002). Even if they didn't—for example, if high-value callers received preferential queuing—it would be difficult for callers to perceive this violation of social justice. Other findings from studies of waiting for service (Durrande-Moreau, 1999; Unzicker, 1999) suggest:

- The longer the duration, the more negative the wait.
- Unoccupied time feels longer than occupied time.
- Anxiety makes waits seem longer.
- Unexpected waits seem longer than known waits.
- Unexplained waits seem longer than explained waits.

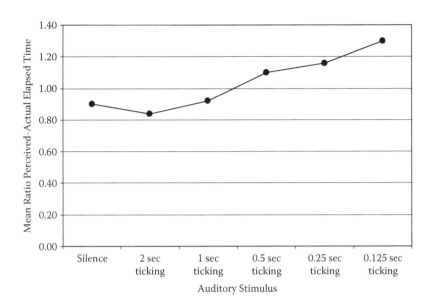

FIGURE 4.4
Ratio of perceived–actual waiting time as a function of ticking rate.

Given an expectation of some waiting time for service, a consideration for the design of service is how to fill that time. For IVRs, the only way to fill the time is with sounds, which also serves the purpose of letting callers know that the call is still connected. Those sounds could be simple processing time tones (such as clock ticks), messages, or music.

Polkosky and Lewis (2002) investigated the effect of ticking rate on perceived wait duration (with wait durations set to last from 3 to 18 seconds) and perceived negative affect (anxiety, stress, and impatience). As shown in Figure 4.4, relatively fast ticking rates (equal to or faster than a half second between ticks) increased the perceived waiting time relative to silence. Ticks occurring 1 second apart had no effect on perceived wait time, and 2-second ticking reduced the perceived waiting time.

The 2-second ticking rate was the most effective for the purpose of reducing the perceived passage of time. Unfortunately, the 2-second rate produced poorer affective ratings than the 1-second rate or silence. Also, although ticking accomplishes the goal of letting a caller know that the call has not dropped, it is likely to become annoying if the wait is several minutes in duration. Polkosky (2001) reported that listeners significantly preferred music to other simpler auditory time fillers such as ticking, even for wait times as short as 5 seconds.

In a set of experiments by Fröhlich (2005), participants found wait times more tolerable if they received confirmation of user input ("OK"), and preferred auditory fillers for silence durations longer than 4 seconds. For the auditory fillers they preferred musical sequences (such as instrumental grooves or drum loops) to natural sounds (clicking clock—presumably a 1-second rate, pouring water) or a synthetic ticking sound (90 beats per minute—a ticking rate of 0.75 seconds).

Music has a long history of use when waiting for service, not only for filling time but also for manipulating mood (Bruner, 1990; Garlin & Owen, 2006; Hui, Dube, & Chebat, 1997). Of particular interest to the goal of making wait times more tolerable is whether music increases positive effect and decreases the perceived passage of time. A related question is the effect of verbal messages during the waiting period. Table 4.8 provides descriptions of key research on these topics. The results are not perfectly consistent, but the major findings appear to be:

- Music usually increases rather than decreases the magnitude of perceived wait times.

- Despite this, customers prefer to hear music while waiting, and the presence of music (especially pleasurable music) increases affective response to the wait and to the service organization.

- Interrupting music to play a simple apology message does not appear to be effective in either reducing perceived wait time or increasing customer satisfaction.

- Playing messages that provide information about the caller's position in the queue or the remaining wait time do not affect perceived wait time but do increase customer satisfaction and reduce abandonment rates.

Clearly, it is better from the customer's point of view to have shorter rather than longer hold times, and enterprises should staff their call centers appropriately to handle their call volumes. When callers must experience a hold time, the system should ideally provide pleasurable music and information that enhances the caller's sense of progress in the queue. There is some evidence (Knott et al., 2003; Knott et al., 2004; Kortum et al., 2008) that giving callers control during the waiting period by offering a choice of music type or access to an interactive voice portal can enhance caller satisfaction, although one must seriously question the overall effectiveness of a real call center with hold times long enough for callers to engage with an interactive voice portal, and given the slight effect of the manipulation of choice in the second experiment of Kortum et al. (2008), the cost of such a strategy might exceed the benefit. Given the lack of comprehensive scientific study of the effect of music and messages, designers of this part of the caller's experience must make certain reasoned decisions and exercise some caution in generalizing from the published research to practice.

For example, with the exception of Ramos (1993), the participants in the various experiments described in Table 4.8 did not wait for service in the same way that real callers to real enterprises wait, and their experiences were universally low cost (Cameron et al., 2003). Because one of the key drivers of

TABLE 4.8

Affective and Time-Perception Effects of Music and Messages while Waiting

Study	Description and Key Findings
Kellaris & Kent, 1992	150 business students with a preference for pop/rock music listened to original digital compositions that had similar characteristics other than modality (major, minor, atonal). Participants judged the lengths and pleasantness of the segments. They were not waiting for service or imagining waiting for service. Participants preferred the segments in the major key, then minor, then atonal. The accuracy of their time judgments was best for atonal (18% over), then minor (23% over), then major (38% over). The actual durations were 2.5 min; the perceived durations, respectively, were 2.95, 3.07, and 3.45 minutes.
Ramos, 1993	The type of hold music played for callers to a Florida protective services abuse hotline was systematically varied each week over a 10-week period using five music styles: classical, popular, music arranged for relaxation, country, and jazz. Across the 10 weeks of the study, the type of music appeared to affect the number of lost calls, with the fewest lost calls in descending order for jazz, country, classical, popular, and relaxation. When reanalyzed as a function of week (see Figure 4.5), there is evidence that call center variables other than music probably reduced the lost call rate starting in the fourth week, resulting in a spuriously high average for classical, popular, and relaxation music, which brings the reported conclusions for this study into doubt with regard to the effects of different musical styles.
Hui, Dube, & Chebat, 1997	116 business students watched a video that included one of four musical extracts of low familiarity and a range of tempos or silence. On an independent rating of pleasure, two extracts received high ratings and two received low ratings. The setting of the video was a branch of a real bank, from the viewpoint of someone next in line for service. Participants imagined that they were waiting in line. After an actual wait of 4 minutes, they estimated the duration of the wait. Differences in tempo had no effect on estimated wait duration. For the high-pleasure extracts, the estimated duration was about 7 minutes; for low-pleasure extracts, about 6 minutes, and for no music, about 5 minutes—all overestimating the actual duration. Despite the longer perceived durations in the presence of high-pleasure music, this condition led to the most positive affective response toward the wait and toward the service organization.
North & Hargreaves, 1999	100 college students were left alone to wait for an experiment to start. The waiting area had an acoustic environment of low, moderate, or high-complexity New Age music (5-minute segments looped for continuous play) or no music (25 students per condition). The main experimental measurement was the number of students leaving before the passage of 20 minutes of waiting time. The complexity of the music did not have any effect (respectively, 4, 3, 4 students leaving), but the absence of music did (12 students leaving). Participants reported less enjoyment of the high-complexity music, and underestimated their actual waiting time of 20 minutes in all conditions (no music: estimation of 16.2 minutes; low: 14.8 minutes; moderate: 14.9 minutes; high: 12.1 minutes).

continued

TABLE 4.8 (continued)

Affective and Time-Perception Effects of Music and Messages while Waiting

Study	Description and Key Findings
North, Hargreaves, & McKendrick, 1999	103 callers responding to an advertisement about completing a survey to receive 5 pounds payment were placed on hold for up to 5 minutes. If a participant did not hang up in less than 5 minutes, the system disconnected the call, so the longest possible hold time was 5 minutes. Before going on hold, they heard a "please hold" message. After going on hold, they experienced one of three on-hold conditions: a simple apologetic message played every 10 seconds ("I'm sorry. The line is busy. Please hold."); three Beatles songs (original); or three Beatles songs (pan pipes version). Within a few seconds of disconnection, a researcher called back to get information about the caller's waiting experience. The mean time before disconnection was shortest for the messages condition (198 seconds), next for the original Beatles (229.5 seconds), and longest for the pan pipes (257 seconds). The different auditory stimuli did not appear to affect callers' estimates of the duration of time that they held. Participants significantly preferred music to messages, and rated the pan pipes music as most closely matching their expectation for hold music.
Guéguen & Jacob, 2002	68 college students placing calls to an educational service had to wait on hold for 5 minutes. There were two hold conditions, with and without music (with the music selected by an independent group of students who rated pleasantness and suitability for use as hold music). Both hold conditions included a message that played every 20 seconds, presumably asking the caller to continue holding. Participants provided estimates of how long they were on hold and how long they would have waited before hanging up if not in an experiment. Participants in both hold conditions overestimated the duration of the wait, with the estimate in the music condition (5.26 min) significantly less than in the no-music condition (6.94 min). The estimation of time the participant would have waited before hanging up was significantly greater in the music condition (2.44 min) than in the no-music condition (2.04 min).
Cameron, Baker, Peterson, & Braunsberger, 2003	127 undergraduate students were left in groups of 8 to 16 people waiting for an experiment to start. While waiting 10 minutes of Bach's "Brandenburg Concerto No. 2" played in the background. After the experiment participants completed a questionnaire. The results of structural equation modeling indicated that music exhibited both cognitive (on wait-length evaluation) and affective (on mood) influences, but that its positive contribution to the overall experience was primarily through mood.
Knott, Pasquale, Miller, Mills, & Joseph, 2003	In a pretest of preference for hold-queue content type, 64 participants, after hearing descriptions of the various content types, indicated preference (in descending order) for music, interactive voice portal (for interactive retrieval of news and other information), entertainment, infotainment, advertisements, and silence. A month later 60 of these participants (none of whom had watches at the time) placed four calls to a simulated call center, ostensibly to check the network quality, and were then on hold for 60, 120, 240, or 480 seconds with the hold time filled with music, an interactive voice

TABLE 4.8 (continued)

Affective and Time-Perception Effects of Music and Messages while Waiting

Study	Description and Key Findings
Knott, Pasquale, Miller, Mills, & Joseph, 2003 (continued)	portal, advertisements, or silence (counterbalanced orders of presentation). Satisfaction ratings differed at all levels of hold time and were significantly greater for the interactive voice portal, then music/ads, then silence. Participants did not significantly overestimate hold times for the interactive portal and music, but did significantly overestimate the time for advertisements and silence.
Knott, Kortum, Bushey, & Bias, 2004	In a follow-up study to Knott et al. (2003), the experimenters manipulated the ability to choose a type of hold music and the length of the duration of the initial hold message. Twenty-four participants (none of whom had watches at the time) placed 16 calls (16 different short IVR tasks) to a simulated call center, experiencing either a short (11 s) or long (21 s) hold queue announcement that either did or did not include a menu for selecting one of four types of music (classical, country, classic rock, or jazz), with wait times of 30, 60, 120, or 240 seconds. After each call they estimated the hold time and completed a satisfaction questionnaire. The actual hold time had a highly significant effect on satisfaction ratings. Satisfaction was higher for choice compared to no-choice conditions, but was not significantly affected by announcement duration. The estimates of duration tended to be overestimates, unaffected by choice but affected by announcement duration.
Kortum & Peres, 2006	40 college students placed calls to a call center and spent time on hold. While on hold they heard an entire song from the Billboard Top 40 hits of the previous 30 years, with songs matched to different hold times and starting with an announcement: "Thank you for calling. All operators are currently assisting other customers. When the following song is over a service representative will be available to help you." After each call, participants estimated the time they had spent on hold and answered questions about familiarity with the song and their satisfaction with the call. Participants tended to overestimate the duration of the waits by about 50% but provided positive ratings of their satisfaction with the music heard on hold.
Whiting & Donthu, 2006	224 students in undergraduate marketing classes provided retrospective ratings about a recent voice-to-voice encounter in which they spent time on hold. The data showed significant correlations between satisfaction and perceived wait time and between music valence (the extent to which the caller liked the music) and wait time. Information cues (messages about estimated wait time, number of people in queue before caller) did not affect participants' perceived wait times (but a limitation of the retrospective approach is that the actual wait times were unknown). A replication in a real call center (211 participants) with experimental manipulation of information cues and presence of music produced virtually identical results.
Munichor & Rafaeli, 2007	123 individuals called to sign up to participate in university experiments. They received random assignment to one of three conditions: apologies, queue location information, or music (Richard Clayderman's "Ballade pour Adeline"), then experienced a 108 second hold. In the conditions

continued

TABLE 4.8 (continued)

Affective and Time-Perception Effects of Music and Messages while Waiting

Study	Description and Key Findings
Munichor & Rafaeli, 2007 (continued)	that included stimuli other than music, the music stopped three times at equally spaced intervals for the presentation of the apology or location message. The apology message was, "We are sorry to keep you waiting. Please hold and you will be answered according to your position in line." The location message was, "You are <X> in line," where <X> is the position in the queue. Abandonment rate was lowest (36%) and satisfaction highest in the location condition. Abandonment rates for music and apologies were 69% and 67%, respectively. In a follow-up study the key dependent variables were perceived waiting time and sense of progress in the queue, with participants placing calls to a fictitious call center then holding. Participants pressed a button to indicate the time at which they would normally have disconnected, but stayed on hold for full 108 seconds so everyone provided an estimate of the same hold duration. The mean perceived waiting time was 153 seconds. The patterns of abandonment and satisfaction in the second study were similar to those in the first. Regression modeling revealed that the best predictor of abandonment rate and satisfaction was the caller's sense of progress in the queue.
Kortum, Bias, Knott, & Bushey, 2008	The first experiment in this paper describes the experiment already summarized above for Knott et al. (2004) in which callers experienced having or not having a choice of musical style, with wait times of 30, 60, 120, or 240 seconds. The second experiment replicated the first, but with a between-subjects rather than a within-subjects design to enhance the generalization of results outside of the laboratory. Each of 128 participants placed one call to a test call center, informed in advance that they would probably have to be on hold for some period of time. As in Knott et al. (2004) the actual hold time had a highly significant effect on satisfaction ratings. The mean satisfaction rating was slightly higher for the choice condition (4.61 for choice, 4.16 for no choice, on a 7-point scale), but the difference was of marginal statistical significance ($p = .06$). There was also no significant effect of choice on estimation of wait duration. Across the four wait durations the average overestimation for the choice condition was 16.3 seconds; for no-choice, it was 17.1 seconds, a difference of less than a second.

satisfaction with SSTs is the extent to which they satisfy an intense need (Meuter et al., 2000), future research would benefit from creating or investigating caller behavior and attitude during higher-cost waiting experiences.

Ramos (1993) illustrates the dangers of simply accepting the conclusions of a widely cited experiment without critically examining the experimental design and results. As summarized in Table 4.8, Ramos had access to the hold music played for a protective services abuse hotline in Florida. She rotated five types of music over a 10-week period and measured the percentage of lost (abandoned) calls for each day of the week, Monday through Sunday, so each type of music had two weeks of measurement. The results of an analysis of variance over all 10 weeks of data indicated a significant difference among

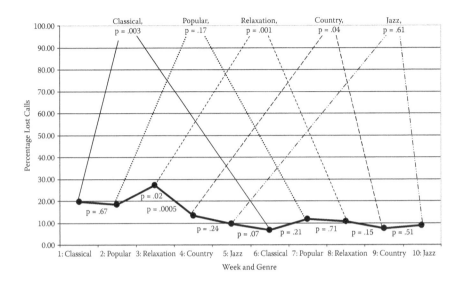

FIGURE 4.5
Data from Ramos (1993) as a function of week.

the music types, with jazz and country having significantly fewer lost calls than relaxation music. On closer examination, however, it appears that one or more other call center variables must have changed during the 10-week course of the study (see Figure 4.5).

The lost call rate for the first 3 weeks hovered around 20%, but for the next 7 weeks hovered around 10%. From week to week, the only significant changes were the increase from Week 2 to Week 3 (t(6) = -3.03, p = 0.02) and the decrease from Week 3 to Week 4 (t(6) = 6.90, p = 0.0005). Comparing the first and second weeks for each musical style shows that for each style its second week had a lower lost call rate than its first week, and for three of the styles (classical, relaxation, and country) the reduction was statistically significant. An analysis of variance restricted to the first set of data for each musical style showed a highly significant difference as a function of week (F(4, 24) = 13.0, p = 0.00001), but the same analysis applied to the second set of data showed no significant difference (F(4, 24) = 1.5, p = 0.23). Contrary to Ramos' (1993) interpretation of the data, a more reasonable interpretation is that something in addition to the change in music happened in the call center between the third and fourth weeks that dramatically reduced the lost call rate. The manipulation of musical style probably did not affect the rate of abandoned calls.

Another limitation of the existing research on user behavior while on hold is the paucity of knowledge about what real callers actually do and care about while on hold in the various conditions. Outside of an experimental setting, callers probably find a variety of ways to fill hold times. On a recent project I listened to recordings of callers on hold who were waiting, sometimes for

TABLE 4.9

Top 10 Secondary Time-Filling Activities When on Hold

Secondary Activity	Reported Percentage
1. Talk with others in same room	62
2. Check or compose e-mail	56
3. Watch television	50
4. Do work or homework	45
5. Doodle	45
6. Eat	42
7. Work on the computer	40
8. Instant message	40
9. Read	36
10. Exercise	32

extended periods of time, to get help applying for or receiving state benefits (cash assistance, food stamps, or health coverage). In addition to callers who just listened to (and occasionally complained about) the unvarying hold music, I heard callers engage in conversation with others in the room or watch television (most of these participants called from their homes). Kortum and Peres (2007) interviewed 101 college undergraduates and found that 79% reported engaging in other activities while on hold, with the top 10 secondary activities listed in Table 4.9. Because participants in the study could select multiple activities, the sum of the percentages is greater than 100.

We (as a profession) really don't know how these additional time-filling activities affect customer attitude toward the wait or toward the service enterprise, but most likely they reduce the negative aspects of the wait to some unknown extent because callers in the real world should usually have more control over how they fill the time than do participants in laboratory experiments. Kortum and Peres (2007) reported a high variation in the secondary activities, both within and between callers.

The research data indicate that, consistent with expectation, the hold experience is better when the caller likes the music that plays. In general people waiting for service seem to prefer music in a major key that matches their expectation of how hold music should sound. Without going overboard, enterprises should make some effort to match hold music style to their typical caller. The complexities of music and changing styles will keep the selection of hold music more of an art than a science. Also, if there is any possibility that callers will experience unusually long hold times, there should be some stylistic variation in the hold music. I know of no published studies on the topic, but any short piece of music (say, 3 minutes) looped continuously for 15–20 minutes is likely to be torturous to listeners and will only exacerbate the negative affective consequences of a long hold.

Regarding messages interspersed in hold music, callers are likely to find commercials irritating, interpreting them as holding up progress while the advertisement plays (Fröhlich, 2005). Simple apologies appear to provide at best no measurable benefit, and at worst may annoy callers (North et al., 1999). For example, each message potentially tricks the caller into thinking the hold is over, only to immediately disappoint the caller when this turns out not to be true. Thus although it is a common practice, enterprises should avoid playing simple apology messages to callers on hold. On the other hand, verbal messages that provide a sense of progress in the queue appear to be effective in reducing abandonment rate and increasing customer satisfaction. As far as I know there is no research on optimal timing of such messages.

Several researchers have experimented with auditory progress bars (APBs)—"tonal cues that continuously convey information about the time left in the hold queue" (Kortum, Peres, Knott, & Bushey, 2005, p. 628). APBs use, for example, variation in volume or pitch to convey progress (Fröhlich, 2005; Kortum et al., 2005). Although with certain configurations, APBs appear to favorably affect the perceived duration of the wait, caller reaction to these types of auditory stimuli are strongly negative (Fröhlich, 2005; Kortum et al., 2005), making them unsuitable for use in an IVR. Kortum and Peres (2006) studied an interesting alternative to APBs, matching known musical selections (Billboard Top 40) to estimated wait times so the song played from start to finish during the wait time. Participants in that experiment estimated wait times as accurately as the better APBs (about 50% overestimation), and rated satisfaction with music as a filler significantly better than any of the tonal APBs.

There are two ways to provide information to a caller about progress in the queue when using verbal messages—the number of people in front of the caller and the time remaining before service. No published experiments have yet investigated the relative effectiveness of these two types of messages, but rational analysis suggests that the better information to offer is the time remaining before service. To translate the number of people before you in a queue into time remaining, you'd have to have some idea of how many service representatives are available to take calls. Customers have this information in physical settings such as waiting in line at a bank or post office, but not when they are on hold. Furthermore, in addition to providing information about progress toward service, knowledge of the estimated time remaining can help callers decide how to fill the remaining time. For example, callers who expect to be on hold for 5 minutes can fill that time with more time-consuming activities (such as reviewing e-mail) than those who expect to be on hold for 30 seconds.

Service Recovery

Successful service recovery takes place when an enterprise overcomes a service failure. As I wrote this chapter I was at a resort in the Bahamas for

New Year's Day 2010. On New Year's Eve, I stood in line at an open bar to get a drink for myself and my wife. I was in line for 15 minutes, and placed my order at 7:03 p.m. The bartender informed me that the open bar closed at 7:00 p.m., but I could charge the drinks to my room. I declined and left with a poor sense of the resort's customer service. A few minutes later a representative from the resort stopped at our table and heard the story about the service failure. Within 10 minutes, we had our drinks free of charge. Through the actions of their representative, the resort achieved a successful service recovery.

Successful service recovery can lead to a phenomenon known as the service recovery paradox. The service recovery paradox occurs when postfailure customer satisfaction exceeds prefailure satisfaction. A recent meta-analysis of 21 empirical research studies on the service recovery paradox (de Matos, Henrique, & Rossi, 2007) led to the following conclusions:

- Despite inconsistencies in the published literature, the service recovery paradox appears to be a real effect with regard to customer satisfaction, at least for the first service failure that a customer experiences with an enterprise.
- Service recovery does not appear to have an effect on customers' repurchase intentions.

After a service failure, showing concern, providing a sizable compensation, and giving customers a way to express their views leads to more positive attitudes than when this does not happen, with females generally caring more than males about having the opportunity to voice complaints (McColl-Kennedy, Daus, & Sparks, 2003). Giving customers a choice among a set of recovery alternatives promotes a more satisfactory recovery than dictating the recovery option to the customer (Chang, 2006). Another way to increase customer satisfaction with service recovery is to give them recovery voice—"asking a customer what the firm can do to rectify a problem after a complaint is voiced" (Karande, Magnini, & Tam, 2007). There is no published research directly comparing the relative effects of offering a set of choices versus a more open-ended request for what recovery action would satisfy a customer.

As discussed previously, apprehension about adequate service recovery in the event of a service failure is a key inhibitor of self-service adoption for both enterprises and customers (Beatson et al., 2007). Unfortunately, for many SST applications the enterprise has inadequate (or nonexistent) plans for service recovery (Bitner et al., 2002). "Planning for service recovery and providing customers with options will help to minimize the negative impact of a failure when one does occur. One difficulty with recovery for SST failures is that SSTs are commonly used where no employees are present to rectify a failure" (Bitner et al., 2002, p. 105). For SSTs in general, they recommend use of a toll-free telephone number, knowledgeable responses to e-mail, or continuous availability of live online chat. Clearly, these are not solutions to the problem of service recovery when the service problem occurs during the use of an IVR.

Given the possibility of a service failure during self-service with an IVR, service recovery strategies include:

- Transfer to a live service representative (if available).
- Enabling a self-service call-back feature (placing the burden of re-establishing the connection on the enterprise).
- Providing information about when live service representatives will be available (placing the burden of re-establishing the connection on the caller).

Consequences of Forced Use of SSTs

Because the cost to an enterprise for transactions completed with an SST is so much less expensive than transactions that involve human employees, it can be tempting to try to force customers to use self-service. However, the strategy of saving money through forced self-service can backfire (Leppik, 2005). Customers expect self-service options that are fast and simple, and they expect to have the choice between self-service and full service. Howard and Worboys (2003), in a study of IVR and Internet self-service, reported that 51% of people claim to prefer dealing with a person at every stage of interaction with an enterprise, and only 5% want self-service at every stage. Only 52% of customers report being satisfied with their self-service experiences, and only 18% of customers stated that IVRs met their needs (Alcock & Millard, 2006).

Furthermore, people have a tendency, known as the status quo bias, to prefer the situation or decision currently in place, even in the face of a potentially better alternative (Falk, Schepers, Hammerschmidt, & Bauer, 2007). There is a reluctance to switch from a service channel known to lead to desired outcomes (for example, speaking to a live representative) to a different service channel for which there is a risk that it will not produce a satisfactory result (for example, a new self-service function in an IVR). Enterprises that take steps to aggressively push customers into using new SSTs could well push their customers into the arms of their competitors. In a survey of more than 1000 railway customers on the topic of choice in selecting service channels, Reinders, Dabholkar, and Frambach (2008) found:

- Forcing customers to use SSTs has adverse effects on attitude toward the SST and toward the service provider.
- Negative attitudes resulting from forced use of SSTs lead to a reduction in positive word of mouth and increased switching intention.

- Offering interaction with an employee as a fallback option offsets the negative consequences of forcing users to initiate service with SSTs.

Thus it is important for most enterprises (especially those who face competition for customers) to make it easy for callers to reach customer service representatives (Larson, 2005). As Balentine (2007, p. 360) stated, "Users who want to speak to an agent will find a way to do it. They will either experiment with different keys (starting with zero), or they will hang up and call back on a different number. Once they reach an agent, they will vent their frustration. If you don't believe me, go ask your agents. Then stand back." And adding injury to insult, the time callers spend venting their frustration increases the time and cost required to complete the call.

From an analysis of 60 IVR studies across mobile service, airline, and financial service industries, Leppik (2005) found that making it harder to reach an agent:

- Slightly increases the number of self-service transactions.
- Causes a strong increase in customer ratings of frustration.
- Causes a dramatic drop in customer satisfaction.

The Leppik (2005) findings also indicated that the simple fact of automation was not the problem. Comparisons of satisfaction and completion with automation showed no significant relationships. Leppik concluded from the data that it is possible to create self-service that works as well as live service, as long as the SST is a well-designed system built with an understanding of the limitations of automation. "The data are very clear: making it hard to reach an agent has only a slight effect on automation rate when you take multiple calls into account, but the price in terms of lower satisfaction and reduced single-call completion is very high" (p. 3).

Summary

Because the primary use of IVRs is to provide self-service, either self-service call routing or more complex self-service functions, the design of IVRs can benefit from an understanding of service science. Implementation of new services generally starts with a focus on talent (human–human interaction) then, as some tasks associated with the service become well-understood and routine, shifts to a focus on automation (self-service technology through human–computer interaction). Because people have a tendency to keep using familiar service channels, it can be difficult to get them to switch from full service to self service.

These general patterns of service evolution appear in the history of telephone-based service, from general operators (high cost) to call centers to skill-based routing to automation (least cost). The characteristics of call centers and effective call center agents can inform the design of IVRs. Highly scripted call center tasks are good candidates for self-service. Even if it is not possible to completely automate a task, there can be significant savings by automating the initial steps. When possible, IVRs should mimic the characteristics of effective call center agents, with scripting that is initially efficient, helpful when the caller appears to need assistance, but not so terse as to appear to be rude.

Studies of the TAM and TRI provide some information about the factors that affect technology adoption. Actual use of technologies is affected by the intention to use, which is itself affected by the perceived usefulness and usability of the technology. There are individual differences that also affect people's likely adoption of a technology, based on differences in attitudinal levels of Optimism, Innovativeness, Discomfort, and Insecurity. Based on the different patterns of these attitudes, researchers have defined customer groups that differ in their adoption of technologies: Explorers, Skeptics, Pioneers, and Laggards. These lines of research indicate that it is important to develop efficient self-service technologies and that it is unrealistic to expect extremely high adoption of technologies. Companies that provide service to primarily male, educated, younger and high-income customers can anticipate relatively higher adoption of their SSTs. Enterprises should continuously evaluate their service strategies to get the right balance between live service and the available array of SSTs.

Studies of SSTs reveal that customers are satisfied with them when they are better than other alternatives, do what they purport to do, and solve an intense need. The main drivers of dissatisfaction are technology failure, poor design, process failure (especially service recovery failure), and customer-driven failure. For successful implementation of SSTs, enterprises should be clear on the purpose of the SST, maintain a customer focus, actively promote the SST, prevent and manage failures, offer choices to customers (including the possibility of live service), and plan for continuous updating and improvement of SSTs.

Customers must sometimes wait for service. Research in waiting, including being on hold for service over the telephone, indicates that it is important to provide music as part of the waiting experience. Music does not appear to reliably reduce the perception of the duration of passing time, but does appear to increase the pleasantness of the wait, which in turn increases caller satisfaction with the wait and attitude toward the service provider. The addition of simple apology messages during the wait does not appear to improve the caller experience, and runs some risk of irritating callers. On the other hand, messages that provide information about progress in the service queue enhance customer satisfaction and can reduce abandonment rates.

Service systems, whether human or automated, sometimes fail. When they fail, it is important to provide for effective service recovery. When service recovery is excellent, the service recovery paradox can occur, with post-failure satisfaction with service being higher than prefailure satisfaction. Excellent service recovery includes an opportunity for the customer to voice the complaint, an expression of concern from the service enterprise, and compensation to the customer. Customers are most satisfied with the compensation they receive when they have a choice, either from a set of alternatives offered by the service representative or as the result of an open-ended request for what recovery action would satisfy a customer. When service failures occur during interaction with an IVR, service recovery strategies include transfer to a live service representative (if available), enabling a self-service call-back feature, or at least providing information about when live service representatives will be available.

Due to the relatively low cost of SSTs it can be tempting to force customers to use them, or to make it very difficult to get to live service. Because some customers will resist the use of SSTs, this strategy can lead to poor customer satisfaction with the SST and with the service provider. The negative attitudes that result from forcing customers to use SSTs lead to a reduction in positive word of mouth and increased switching intention. Offering interaction with an employee as a fallback option can offset the negative consequences of forced use of SSTs. Making it hard for callers to reach live service has only a slight effect on increasing automation rate and in return extols a high price in terms of lower satisfaction and reduced single-call completion.

References

Alcock, T., & Millard, N. (2006). Self-service—but is it good to talk? *BT Technology Journal, 24*(1), 70–78.

Armistead, C., Kiely, J., Hole, L., & Prescott, J. (2002). An exploration of managerial issues in call centres. *Managing Service Quality, 12*(4), 246–256.

Balentine, B. (2007). *It's better to be a good machine than a bad person.* Annapolis, MD: ICMI Press.

Balentine, B., & Morgan, D. P. (2001). *How to build a speech recognition application: A style guide for telephony dialogues* (2nd ed.). San Ramon, CA: EIG Press.

Beatson, A., Lee, N., & Coote, L. V. (2007). Self-service technology and the service encounter. *The Service Industries Journal, 27*(1), 75–89.

Beckett, A. (2004). From branches to call centres: New strategic realities in retail banking. *The Service Industries Journal, 24*(3), 43–62.

Bergevin, R., & Wyatt, A. (2008). *Contact centers for dummies.* Indianapolis, IN: Wiley Publishing.

Bitner, M. J., Ostrom, A. L., & Meuter, M. L. (2002). Implementing successful self-service technologies. *Academy of Management Executive, 16*(4), 96–108.

Bruner, G. C. II. (1990). Music, mood, and marketing. *Journal of Marketing, 54*(4), 94–104.

Cameron, M. A., Baker, J., Peterson, M., & Braunsberger, K. (2003). The effects of music, wait-length evaluation, and mood on a low-cost wait experience. *Journal of Business Research, 56,* 421–430.

Chang, C. (2006). When service fails: The role of the salesperson and the customer. *Psychology & Marketing, 23*(3), 203–224.

Cunningham, L. F., Young, C. E., & Gerlach, J. H. (2008). Consumer views of self-service technologies. *The Service Industries Journal, 28*(6), 719–732.

Curran, J. M., & Meuter, M. L. (2005). Self-service technology adoption: Comparing three technologies. *Journal of Service Marketing, 19*(2), 103–113.

Curran, J. M., Meuter, M. L., & Surprenant, C. F. (2003). Intentions to use self-service technologies: A confluence of multiple attitudes. *Journal of Service Research, 5*(3), 209–224.

Davis, D. (1989). Perceived usefulness, perceived ease of use, and user acceptance of information technology. *MIS Quarterly, 13*(3), 319–339.

de Matos, C. A., Henrique, J. L., & Rossi, C. A. V. (2007). Service recovery paradox: A meta-analysis. *Journal of Service Research, 10*(1), 60–77.

Durrande-Moreau, A. (1999). Waiting for service: Ten years of empirical research. *International Journal of Service Industry Management, 10*(2), 171–189.

Falk, T., Schepers, J., Hammerschmidt, M., & Bauer, H. H. (2007). Identifying cross-channel dissynergies for multichannel service providers. *Journal of Service Research, 10*(2), 143–160.

Fluss, D. (2008). Are you doing right by your customers? *Speech Technology, 13*(8), 38–39.

Fröhlich, P. (2005). Dealing with system response times in interactive speech applications. In *Proceedings of CHI 2005* (pp. 1379–1382). Portland, OR: ACM.

Garlin, F. V., & Owen, K. (2006). Setting the tone with the tune: A meta-analytic review of the effects of background music in retail settings. *Journal of Business Research, 59,* 755–764.

Guéguen, N., & Jacob, C. (2002). The influence of music on temporal perceptions in an on-hold waiting situation. *Psychology of Music, 30,* 210–214.

Howard, M., & Worboys, C. (2003). Self-service—a contradiction in terms or customer-led choice? *Journal of Consumer Behaviour, 2*(4), 382–392.

Hui, M. K., Dube, L., & Chebat, J. (1997). The impact of music on consumers' reactions to waiting for services. *Journal of Retailing, 73*(1), 87–104.

Karande, K., Magnini, V. P., & Tam, L. (2007). Recovery voice and satisfaction after service failure: An experimental investigation of mediating and moderating factors. *Journal of Service Research, 10*(2), 187–203.

Kellaris, J. J., & Kent, R. J. (1992). The influence of music on consumers' temporal perceptions: Does time fly when you're having fun? *Journal of Consumer Psychology, 1*(4), 365–376.

Kleijnen, M., de Ruyter, K., & Wetzels, M. (2007). An assessment of value creation in mobile service delivery and the moderating role of time consciousness. *Journal of Retailing, 83*(1), 33–46.

Knott, B. A., Kortum, P., Bushey, R. R., & Bias, R. (2004). The effect of music choice and announcement duration on subjective wait time for call center hold queues. In *Proceedings of the Human Factors and Ergonomics Society 48th annual meeting* (pp. 740–744). Santa Monica, CA: HFES.

Knott, B. A., Pasquale, T., Miller, J., Mills, S. H., & Joseph, K. M. (2003). "Please hold for the next available agent:" The effect of hold queue content on apparent hold duration. In *Proceedings of the Human Factors and Ergonomics Society 47th annual meeting* (pp. 668–672). Santa Monica, CA: HFES.

Kortum, P., Bias, R. G., Knott, B. A., & Bushey, R. G. (2008). The effect of choice and announcement duration on the estimation of telephone hold time. *International Journal of Technology and Human Interaction, 4*(4), 29–53.

Kortum, P., & Peres, S. C. (2006). An exploration of the use of complete songs as auditory progress bars. In *Proceedings of the Human Factors and Ergonomics Society 50th annual meeting* (pp. 2071–2075). Santa Monica, CA: HFES.

Kortum, P., & Peres, S. C. (2007). A survey of secondary activities of telephone callers who are put on hold. In *Proceedings of the Human Factors and Ergonomics Society 51st annual Meeting* (pp. 1153–1157). Santa Monica, CA: HFES.

Kortum, P., Peres, S. C., Knott, B. A., & Bushey, R. (2005). The effect of auditory progress bars on consumer's estimation of telephone wait time. In *Proceedings of the Human Factors and Ergonomics Society 49th annual meeting* (pp. 628–632). Santa Monica, CA: HFES.

Larson, J. A. (2005). Ten guidelines for designing a successful voice user interface. *Speech Technology, 10*(1), 51–53.

Larson, R. C. (1987). Perspectives on queues: Social justice and the psychology of queuing. *Operations Research, 35*(6), 895–905.

Larson, R. C. (2008). Service science: At the intersection of management, social, and engineering sciences. *IBM Systems Journal, 47*(1), 41–51.

Leppik, P. (2005). Does forcing callers to use self-service work? *Quality Times, 22,* 1–3. Retrieved from http://www.vocalabs.com/resources/newsletter/newsletter22.html

Liljander, V., Gillberg, F., Gummerus, J., & van Riel, A. (2006). Technology readiness and the evaluation and adoption of self-service technologies. *Journal of Retailing and Consumer Services, 13,* 177–191.

Lin, J. C., & Hsieh, P. (2007). The influence of technology readiness on satisfaction and behavioral intentions toward self-service technologies. *Computers in Human Behavior, 23,* 1597–1615.

Lusch, R. F., Vargo, S. L., & O'Brien, M. (2007). Competing through service: Insights from service-dominant logic. *Journal of Retailing, 83*(1), 5–18.

Lusch, R. F., Vargo, S. L., & Wessels, G. (2008). Toward a conceptual foundation for service science: Contributions from service-dominant logic. *IBM Systems Journal, 47*(1), 5–14.

McColl-Kennedy, J. R., Daus, C. S., & Sparks, B. A. (2003). The role of gender in reactions to service failure and recovery. *Journal of Service Research, 6*(1), 66–82.

Meuter, M. L., Bitner, M. J., Ostrom, A. L., & Brown, S. W. (2005). Choosing among alternative service delivery modes: An investigation of customer trial of self-service technologies. *Journal of Marketing, 69,* 61–83.

Meuter, M. L., Ostrom, A. L., Bitner, M. J., & Roundtree, R. (2003). The influence of technology anxiety on consumer use and experiences with self-service technologies. *Journal of Business Research, 56,* 899–906.

Meuter, M. L., Ostrom, A. L., Roundtree, R., & Bitner, M. J. (2000). Self-service technologies: Understanding customer satisfaction with technology-based service encounters. *Journal of Marketing, 64,* 50–64.

Munichor, N., & Rafaeli, A. (2007). Numbers or apologies? Customer reactions to telephone waiting time fillers. *Journal of Applied Psychology, 92*(2), 511–518.

North, A. C., & Hargreaves, D. J. (1999). Can music move people? The effects of musical complexity and silence on waiting time. *Environment and Behaviour, 31,* 136–149.

North, A. C., Hargreaves, D. J., & McKendrick, J. (1999). Music and on-hold waiting time. *British Journal of Psychology, 90,* 161–164.

Ostrom, A., Bitner, M., & Meuter, M. (2002). Self-service technologies. In R. Rust & P. K. Kannan, (Eds.), *E-service: New directions in theory and practice* (pp. 45–64). Armonk, NY: M.E. Sharpe.

Osuna, E. E. (1985). The psychological cost of waiting. *Journal of Mathematical Psychology, 29,* 82–105.

Parasuraman, A. (2000). Technology readiness index (TRI): A multiple-item scale to measure readiness to embrace new technologies. *Journal of Service Research, 2*(4), 307–320.

Parasuraman, A., & Colby, C. L. (2001). *Techno-ready marketing: How and why your customers adopt technology.* New York, NY: Free Press.

Pitkänen, O., Virtanen, P., & Kemppinen, J. (2008). Legal research topics in user-centric services. *IBM Systems Journal, 47*(1), 143–152.

Polkosky, M. D. (2001). *User preference for system processing tones* (Tech. Rep. 29.3436). Raleigh, NC: IBM.

Polkosky, M. D., & Lewis, J. R. (2002). Effect of auditory waiting cues on time estimation in speech recognition telephony applications. *International Journal of Human-Computer Interaction, 14,* 423–446.

Rafaeli, A., Barron, G., & Haber, K. (2002). The effects of queue structure on attitudes. *Journal of Service Research, 5*(2), 125–139.

Rafaeli, A., Ziklik, L., & Doucet, L. (2008). The impact of call center employees' customer orientation behaviors on service quality. *Journal of Service Research, 10*(3), 239–255.

Ramos, L. (1993). The effects of on-hold telephone music on the number of premature disconnections to a statewide protective services abuse hot line. *Journal of Music Therapy, 30*(2), 119–129.

Reinders, M., Dabholkar, P. A., & Frambach, R. T. (2008). Consequences of forcing consumers to use technology-based self-service. *Journal of Service Research, 11*(2), 107–123.

Rowley, J. (2006). An analysis of the e-service literature: Towards a research agenda. *Internet Research, 3,* 339–359.

Spohrer, J., & Maglio, P. P. (2008). The emergence of service science: Toward systematic service innovations to accelerate co-creation of value. *Production and Operations Management, 17*(3), 238–246.

Strawderman, L., & Koubek, R. (2008). Human factors and usability in service quality measurement. *Human Factors and Ergonomics in Manufacturing, 18*(4), 454–463.

Tsikriktsis, N. (2004). A technology readiness-based taxonomy of customers: A replication and extension. *Journal of Service Research, 7*(1), 42–52.

Unzicker, D. K. (1999). The psychology of being put on hold: An exploratory study of service quality. *Psychology & Marketing, 16*(4), 327–350.

Whiting, A., & Donthu, N. (2006). Managing voice-to-voice encounters: Reducing the agony of being put on hold. *Journal of Service Research, 8*(3), 234–244.

Wu, J., Chen, Y., & Lin, L. (2007). Empirical evaluation of the revised end user computing acceptance model. *Computers in Human Behavior, 23,* 162–174.

Xue, M., & Harker, P. T. (2002). Customer efficiency: Concept and its impact on e-business management. *Journal of Service Research, 4*(4), 253–267.

Yellin, E. (2009). *Your call is (not that) important to us: Customer service and what it reveals about our world and our lives.* New York, NY: Free Press.

5

The Importance of Speech
User Interface Design

When properly designed using current technologies, SUI applications have tremendous potential for user acceptance, especially relative to their touchtone cousins. To achieve a proper design, however, requires an understanding of SUI design disciplines, an effective SUI design team, and appropriate goals. This chapter provides an overview of the research on user acceptance of speech applications and discussion of the disciplines of SUI design, the consumers of SUI designs, and major SUI objectives.

User Acceptance of Speech IVR Applications

Market Research

One of the most widely cited surveys on user acceptance of speech IVR applications was a Harris Interactive survey conducted in 2003 (Anton & Kowal, 2005; Bailey, 2004; Chiarelli, 2003; Herrell, 2004; Nuance, 2003), which reported the following key results based on interviews with 326 respondents about their most recent interaction with a speech application:

- Speech is widely used and accepted (66% encountered speech applications regularly; only 7% of respondents in the survey would avoid future use of speech systems; 56% indicated an intention to definitely or probably use the speech application again).
- Consumers reported high satisfaction with speech experiences (61% highly satisfied with most recent speech interaction).
- Consumers felt that speech provided many advantages over touchtone (90% of respondents preferred speech to touchtone systems).

Syntellect (2002), a contact center company, published a review of the business case for speech-enabled self-service solutions. It was also very positive regarding user acceptance of speech IVR applications. Some of their key statements were:

- The typical annual cost (labor and burden) of a call center agent was about $36,000, with about 23% turnover.
- Pre-call processing and skill-based routing reduce the average length of calls by 18% and improve call resolution and customer satisfaction.
- 83% of callers prefer speech to touchtone
- Speech recognition can increase IVR usage from 20–60%.
- Relative to touchtone, speech recognition can reduce average call length by 30–50%.

Bailey (2004), in a report for Apex Research, focused specifically on a review of speech versus touchtone in the financial services industry, reporting that:

- Natural language call routing is typically more successful than touchtone in getting callers to their desired call routing destination (82% versus 57%).
- Companies with well-developed touchtone applications can expect a 10–20% improvement in usage after adding speech.
- 80% of callers prefer speech to touchtone applications.
- 70% of the top 10 financial companies (and 56% of the top 25) used speech in their IVR applications, and reported breaking even on the investment in 10 months on average.
- On switching from touchtone to speech, the percentage of satisfied customers increased from 59.8% to 79.0%; abandonment rates decreased from 11.7% to 7.2%; and zero-out rates (requests to transfer to an agent) went down from 21.5% to 16.8%.
- A usability study of 762 callers checking their balance with a financial services company's IVR revealed that 59% rated the application as exceeding their expectations; it fell short of expectation for 17%.

A Datamonitor report on myths and realities of the value of speech in contact centers (Hong, 2004) was similarly positive regarding user acceptance of speech IVR applications, but with the caveat that speech alone cannot compensate for bad user interface design, and that most contact center IT departments do not have the proper staffing to perform high-quality SUI design.

Forrester Consulting (2009) surveyed 1001 online consumers to gauge their interest in automated telephone customer service systems, with a focus on getting information about the situations in which consumers find IVRs to be most useful and valuable. They reported that only 49% of respondents were satisfied, very satisfied, or extremely satisfied with customer service in general. They engaged with automated customer service on a regular basis (82%, second only to 93% contact with live agents). Participants indicated a preference for automated over live agent service for prescription refill, flight status, account balance, and store hours. Preference was about evenly split for tracking shipments and bill pay. Most callers preferred live assistance for health insurance claim status, health provider coverage, upgrading pay TV service, and questioning suspicious bank charges. When asked what they valued about automated customer service, respondents cited full-time availability (77%), no hold time (40%), and quick service (31%). When asked what makes a great experience with an automated speech-enabled customer service system, the top response was the option to speak to a live agent at any time (67%).

VocaLabs (Leppik, 2010), in a national survey of satisfaction with computer technical support, reported that 84% of callers who rated themselves as very satisfied with the automated portion of the service call had a repurchase intention, compared with 49% of customers who were not satisfied with the call. This effect, although not as pronounced as satisfaction with the agent or the call overall, illustrates the importance of good IVR design. The data from this survey highlighted the importance of first-call resolution (86% of callers whose issue was resolved on the first call were very satisfied; 12% of other callers were very satisfied). For the company that improved the most, an 11% decrease in IVR problems, 11% increase in perceived ease of reaching an agent, 7% increase in perception of a reasonable wait time, and 12% decrease in complaints about irrelevant steps led to a 6% increase in overall satisfaction.

Agent and Channel Costs

Table 5.1 summarizes a number of different estimates of the cost of call center agents. The Syntellect (2002) estimate of agent cost of $36,000 per year comes to 30 cents per minute (assuming 2000 work hours per year—because agents do not spend 100% of their work time in calls, this could be a significant underestimate). Hong (2004) used $10 per hour for agent cost and 12 cents per minute for toll costs, which translates to 29 cents per minute (but note that at $10 per hour, Hong's agents have an average annual salary of $20,000 assuming 2000 work hours per year, which is much lower than the Syntellect estimate of $36,000). Kuo, Siohan, and Olive (2003) estimated cost at about $1 per minute of agent time—an estimate consistent with the calculations

TABLE 5.1

Published Estimates of Cost per Minute of Call Center Agents

Source	Cost per Minute
Syntellect, 2002	$0.30
Kuo, Siohan, & Olive, 2003	$1.00
Hong, 2004	$0.29
Anton & Kowal, 2005	$1.07
Suhm, 2008	$0.70

TABLE 5.2

Typical Costs of Issue Resolution by Channel

Channel	Typical Cost
E-mail	$9.53
Chat	$7.86
Web	$6.55
Phone (agent)	$6.17
Phone (speech recognition)	$0.20

of Anton and Kowal (2005), who used an average call time of 4.61 minutes and average call cost of $4.95 ($1.07 per min) in their example of return on investment (ROI). Suhm (2008) estimated agent cost at about 70 cents per minute. Herrell (2004) reported an estimate of the cost of interacting with a speech application as 10 to 25 cents per minute. It seems reasonable to keep back-of-the-envelope ROI estimates simple by using $1 per minute for agent time and 20 cents per minute for speech IVR time, but any real computation of ROI for a company should use its actual expense data.

Table 5.2 shows data reported by Nuance (2005) for the costs of resolving issues as a function of communication channel. The typical cost of calls resolved using a speech recognition application was much lower than that for the other channels. The same report contained data from three case studies in which moving from touchtone to speech substantially improved self-service rates (from 50% with touchtone to 88% with speech for a North American airline, from 45% to 63% for a mutual fund company, and from 82% to 90% for a global financial services firm).

In 2005, Opus Research surveyed companies that had made investments in speech IVR applications (Miller, 2006). They found that 80% of respondents achieved financial results that met or exceeded their objectives. Only 5% did not reduce cost. The remaining 15% achieved lower cost, but did not meet their self-defined objectives. The average payback period on investment was 11 months.

Scientific Research

During the same period of time as these market research papers, similar findings appeared in the scientific literature. Suhm, Bers, McCarthy, Freeman, Getty, Godfrey, & Peterson (2002) built a natural language (NL) call router and conducted a field study in a call center of a large telecommunication provider. Calls were routed to one of five possible destinations: order specialist, repair specialist, billing agent, payment specialist, or default agent. During the time of the study, most callers (95,904) experienced the existing touchtone routing process, but a randomly selected subset of callers (3759) experienced the NL router. In addition to collecting a set of automated performance measures, 945 randomly selected users of the NL call router completed a preference questionnaire. The average routing time for touchtone was 35.9 seconds compared to 16.5 seconds for the NL call router. The misdirection rate (getting to the wrong skill group) was lower with NL call routing: about 12% for NL call routing and 19% for the touchtone menus. Of those callers indicating a preference, 80% preferred speech to touchtone. 94% of callers rated the NL router from "satisfactory" to "outstanding"; 6% as "not so good" or "poor."

Lee and Lai (2005) compared the use of speech and touchtone in a voice-mail message retrieval system. Sixteen participants used both versions of the message retrieval system in counterbalanced order to complete six different tasks. The touchtone method was generally more efficient for simple, linear tasks, but the speech method was more efficient for nonlinear tasks (for example, checking for a message from a particular person). The speech version received significantly better ratings of ease of use than the touchtone version, and 75% of participants chose speech as their preferred modality.

Location on the "Hype Cycle"

Gartner, Inc. periodically reviews the available market and scientific research to assess where human-computer interaction technologies fall on their proposed hype cycle. Based on observations about the life-cycle of new technologies, the five stages of the hype cycle are:

- Technology Trigger: This is the initial event that generates significant press and industry interest in the technology.
- Peak of Inflated Expectations: The primary characteristic of this phase is overenthusiastic promotion (hype) of the technology, leading to some marketplace successes but more failures due to initial limitations, both of the technology and the understanding of how to effectively use the technology.
- Trough of Disillusionment: As a reaction to the former hype and marketplace failures, the technology becomes unpopular, both in the industry and in the press.
- Slope of Enlightenment: In this stage, continuing experimentation and solid hard work with the technology leads to an improved understanding of the technology's applicability, risks, and benefits.
- Plateau of Productivity: The technology, now reasonably well-understood, enters more mainstream usage.

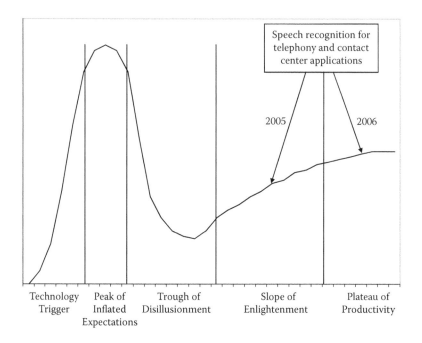

FIGURE 5.1
Location of speech recognition for telephony and call center application on the 2005/2006 Gartner Hype Cycle.

As shown in Figure 5.1, consistent with the evidence from marketing and scientific research in the early to mid-2000s, Gartner moved the location of speech recognition for telephony and call center applications from the middle of the Slope of Enlightenment in 2005 to the Plateau of Productivity in 2006 (Fenn, Cramoysan, Elliot, Bell, Davies, Dulaney, et al., 2006).

Despite the clear potential effectiveness of and preference for speech recognition over touchtone IVRs, speech systems in general have a long way to go to become a preferred contact channel. Recent market research by Convergys (Klie, 2009) found that of 3000 enterprise customers, speaking with a live agent was the preferred customer service option (68% in the United States, 63% in the UK). In the United States, only 4% listed speech-activated menus as their preferred customer service option, with 3% preferring touchtone (in the UK, 2% and 2%, respectively). Second-place preferences were live chat (31% US, 28% UK), company Web sites (22% US, 20% UK), e-mail (17% US, 25% UK), touchtone (11% US, 8% UK), and speech-activated menus (10% US, 6% UK). It is very likely that, in addition to the limitations of the technology, suboptimal SUI design will continue to contribute to reduced customer acceptance of speech recognition IVR applications.

A positive finding for the speech industry was that customers placed a very high value on first-contact resolution. In the United States, 65% of respondents (58% in the UK) would prefer to use an automated solution once to resolve a problem rather than having multiple contacts with a person. In both countries, 55% would rather use an automated solution rather than to wait on hold. The percentages of preference for automation versus live agent vary as a function of the specific task that the caller wants to accomplish (Forrester Consulting, 2009—see "Market Research" above).

The Disciplines of SUI Design and the SUI Design Team

As illustrated in Figure 5.2, effective SUI design draws upon many disciplines. The previous three chapters covered relevant research in three key scientific areas—speech technologies (Chapter 2), psychology and linguistics (Chapter 3), and market research of self-service technologies (Chapter 4). As we move forward through the remaining chapters into the details of designing speech applications, the focus on research-based design will continue, but elements of art and craft will appear where current research does not inform design.

Figure 5.2 also provides information about the requisite skills for a SUI design team. Effective designs require knowledge of system and human

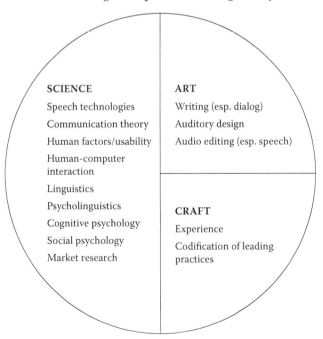

FIGURE 5.2
The disciplines of SUI design.

operating characteristics, plus the ability to translate that knowledge into design, using art and craft to fill in the gaps in the research. There shouldn't necessarily be a one-to-one mapping between the elements in Figure 5.2 and team personnel, but the combined team should adequately cover the elements. Note that this addresses SUI design but does not address other critical aspects of the overall application development team, such as architects and programmers.

The Consumers of SUI Design

In addition to its end users, a SUI design has many other consumers and must simultaneously satisfy many objectives. Among the other consumers of SUIs are marketers, service providers, and developers (see Figure 5.3).

End users call the speech application for the purpose of obtaining a service from the service provider. End users want SUIs that are easy to use, allow efficient task completion, and provide a pleasant user experience.

Marketers have the responsibility of selling speech applications to service providers. The primary objective of a SUI from a marketer's point of view is to appeal to the targeted service provider, thus helping to make the sale.

Service providers rely on the speech application to help them provide a service to end users. For service providers the primary objectives of a SUI are to save money and maintain customer contact and satisfaction. To this end,

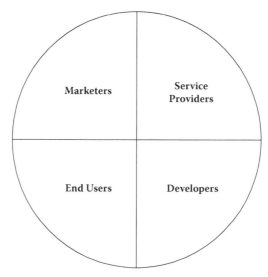

FIGURE 5.3
The consumers of SUI design.

they are also concerned with their corporate image and how their speech applications will fit into their overall branding strategies.

Developers must write the code that creates the entire speech application, including the SUI. The primary objectives for developers creating SUIs are that the interface be technologically feasible, capable of completion given resource constraints, and require minimal development effort.

Major SUI Objectives

The main enemy of the spoken user interface is time (Balentine & Morgan, 2001). Speech has a temporary existence and listeners must remember what they have heard. If prompts in a speech application are too short, however, they can be subject to multiple cognitive and social interpretations. A clear design objective, therefore, is to avoid making users hear more (or less) than they need to hear or to say more (or less) than they need to say.

It is also important to strive to make every interaction move the user forward (or at least create the illusion of moving forward). This is easier said than done because dialogs need careful crafting and usability evaluation. To work toward these objectives:

- Use prompts that are succinct and sincere (modeled after the prompts used by expert call center agents).
- Provide self-revealing contextual help messages.
- Use a professional voice talent for recorded prompts.

The Components of SUI Usability

These goals align with the results of recent investigation into the components of SUI usability (Polkosky, 2005a, 2005b). This research demonstrated that there are four major components of SUI usability, listed in Table 5.3.

In general, greater system adherence to User Goal Orientation, Customer Service Behavior, and Speech Characteristics lead to greater customer satisfaction. The elements associated with Verbosity (a system that talks too much and is too repetitive) lead to poorer customer satisfaction. These findings are consistent with earlier descriptions of the characteristics of effective call center agents (Balentine & Morgan, 2001) and their relationship to the concept of a system persona.

Given the nature of human speech recognition, including the recognition of paralinguistic information as well as the actual content of speech, there's always

TABLE 5.3

The Components of SUI Usability

Component	Description
User Goal Orientation	The efficiency, control, and confidence allowed by the system as users complete tasks
Customer Service Behavior	The user expectations associated with customer service, such as use of everyday words, friendliness, helpfulness, and politeness
Verbosity	System talkativeness, use of repetitive messages, and amount of detail
Speech Characteristics	The enthusiasm, naturalness, and pleasantness of the system voice

a persona implied by an IVR's voice and style. There is, however, no need to spend enormous resources on providing a name or an elaborate back story for the system persona. Instead, design to mimic the important characteristics of a professional call center agent's style in dealing with callers. In particular:

- Assume the caller is busy.
- Be efficient when communication is good.
- Be helpful when progress is slower.
- Be courteous and polite, rarely apologize, and never blame the caller.

The Power of the SUI

A good SUI balances the multiple objectives of User Goal Orientation, Customer Service Behavior, Verbosity, and Speech Characteristics. There is no need to associate a function with a number (in contrast to touchtone user interfaces) when speech labels provide good functional descriptions, which relieves demands on human working memory. System prompts become much shorter and more natural (again reducing demands on human working memory), and it is possible to add options in the future without any need to change existing speech labels, even if the order of functions changes. Finally, there is no need for a user to move the phone away from his or her ear to find the button to press.

Summary

Both market and scientific research support the use of speech recognition in IVR applications, especially in contrast to touchtone applications. The data also indicate that although user acceptance of speech recognition is high, it is not 100%. Numerous studies estimate the preference of speech over touchtone at about 80% (Bailey, 2004; Lee & Lai, 2005; Nuance, 2003; Suhm et al., 2002; Syntellect, 2002). From the full range of customer contact technologies, IVRs are among the less preferred options (Klie, 2009), very likely due to inherent limitations of the associated technologies and to the effects of suboptimal SUI design.

Effective SUI design draws upon multiple disciplines from science, art, and craft. The resulting designs must satisfy a number of different consumers, including end users, marketers, service providers, and developers. To satisfy end users (the ultimate goal for service providers), the designs must address the key components of SUI usability: strong user goal orientation, good customer service behavior, low verbosity, and pleasant speech characteristics (Polkosky, 2005a, 2005b). Mimicking the important characteristics of professional call center agents, speech IVR applications should assume the caller is busy, be efficient when progress is good, be helpful when progress is slower, be courteous and polite, rarely apologize, and never blame the caller (Balentine & Morgan, 2001).

By following these very high-level goals, speech IVR applications also conform to key human factors engineering objectives. Elimination of touchtone number-to-function mapping and efficient use of speech reduce demands on human working memory. While always keeping in mind that human-machine spoken communication is not truly "natural" and likely will not be for many years to come (Balentine, 2007), the use of wording and social patterns familiar to callers takes advantage of population stereotypes and social scripts, which makes it easier for callers to complete tasks successfully with less effort and with higher satisfaction than when they must deal with speech interfaces that do not match expectation.

References

Anton, J., & Kowal, G. P. P. (2005). *Enabling IVR self-service with speech recognition.* Santa Maria, CA: Anton Press.

Bailey, J. (2004). *Speech versus DTMF within the financial services industry.* Seattle, WA: Apex Research.

Balentine, B. (2007). *It's better to be a good machine than a bad person.* Annapolis, MD: ICMI Press.

Balentine, B., & Morgan, D. P. (2001). *How to build a speech recognition application: A style guide for telephony dialogues* (2nd ed.). San Ramon, CA: EIG Press.

Chiarelli, K. (2003). Dial in and speak up. *Speech Technology, 8*(5), 40–41.

Fenn, J., Cramoysan, S., Elliot, B., Bell, T., Davies, J., Dulaney, K., et al. (2006). *Hype cycle for human-computer interaction, 2006.* Stamford, CT: Gartner, Inc.

Forrester Consulting. (2009). *Driving consumer engagement with automated telephone customer service*. Cambridge, MA: Forrester.

Herrell, E. (2004). *Evaluating speech self-service platforms*. Cambridge, MA: Forrester.

Hong, D. (2004). *Proving the value of speech technology in the contact center: Myths vs. reality*. New York, NY: Datamonitor.

Klie, L. (2009). Preference low for voice channels. *Speech Technology, 14*(8), 12.

Kuo, H. J., Siohan, O., & Olive, J. P. (2003). Advances in natural language call routing. *Bell Labs Technical Journal, 7*(4), 155–170.

Lee, K. M., & Lai, J. (2005). Speech versus touch: A comparative study of the use of speech and DTMF keypad for navigation. *International Journal of Human-Computer Interaction, 19*(3), 343–360.

Leppik, P. (2010). *National customer service survey: Computer tech support*. Golden Valley, MN: VocaLabs.

Miller, D. (2006). Revisiting the ROI of speech. *Speech Technology, 11*(1), 12.

Nuance. (2003). *Harris interactive speech satisfaction survey*. Burlington, MA: Nuance.

Nuance. (2005). *Delivering cost savings and the best customer experience: Speech versus DTMF*. Burlington, MA: Nuance.

Polkosky, M. D. (2005a). Toward a social-cognitive psychology of speech technology: Affective responses to speech-based e-service. Unpublished doctoral dissertation. University of South Florida.

Polkosky, M. D. (2005b). What is speech usability, anyway? *Speech Technology, 10*(9), 22–25.

Suhm, B. (2008). IVR usability engineering using guidelines and analyses of end-to-end calls. In D. Gardner-Bonneau & H. E. Blanchard (Eds.), *Human factors and voice interactive systems* (2nd ed.) (pp. 1–41). New York, NY: Springer.

Suhm, B., Bers, J., McCarthy, D., Freeman, B., Getty, D., Godfrey, K., & Peterson, P. (2002). A comparative study of speech in the call center: Natural language call routing vs. touchtone menus. In *Proceedings of CHI 2002* (pp. 283–290). Minneapolis, MN: Association for Computing Machinery.

Syntellect. (2002). *Is your CRM strategy ready for the freedom of speech: The business case for speech-enabled self service solutions*. Phoenix, AZ: Syntellect.

6

Speech User Interface Development Methodology

Developing speech IVR applications, like most development activities, involves a process with multiple phases. Figure 6.1 illustrates a common "waterfall" approach to software development, applied to the development of speech IVR applications.

In contrast to the waterfall model, there are other possible approaches to software development, such as "agile" methods (Larmon & Basili, 2003) and more flexible methods applied by highly experienced voice application

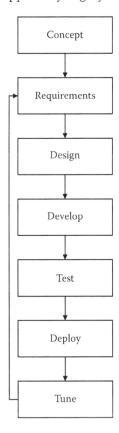

FIGURE 6.1
The SUI development cycle.

development teams (Hura, Turney, Kaiser, Chapin, McTernan, & Nelson, 2008)—for example, in some situations a specific speech application development project might not include one or more steps. That said, the steps in Figure 6.1 capture common elements of current approaches to speech application development (Cohen, Giangola, & Balogh, 2004).

Concept Phase

The impetus to develop an IVR can come from different sources. Inside the enterprise, it is common for the personnel responsible for the call center to continuously monitor the effectiveness of their existing systems, including the analysis of competitors' systems and practices. Vendors who develop IVRs often learn of the desire of an enterprise to invest in their IVR at this stage through a Request for Proposal (RFP).

Requirements Phase

In this phase, the goal is to determine what the application should do. For the SUI designer, this involves analyzing users and their tasks.

Analyzing Users

When developing requirements, the first step in designing a voice application should be to conduct user analysis to identify any user characteristics, perspectives, and requirements that might influence application design (Polkosky, 2006). User characteristics that affect performance with systems include physical, mental, and sensory abilities (Gawron, Drury, Czaja, & Wilkins, 1989). For the use of IVRs, the primary sensory capability is auditory acuity. Key mental abilities for IVR usage are working memory (short-term memory, what a person has just perceived and is currently thinking about), semantic memory (including vocabulary and the rules of language), procedural memory (how to perform activities), attention (awareness of events in the environment), and language comprehension (ability to interpret written or spoken language).

A key physical variable that affects these sensory and cognitive abilities is age (Dulude, 2002; Fisk, Rogers, Charness, Czaja, & Sharit, 2009; Stewart & Wingfield, 2009). As the percentage of users aged over 60 continues to increase worldwide, it is important for IVR designers to know the distribution of ages in their user base. If these data are not available, it is reasonable to design in accordance with the capabilities of older adults. "In most instances, systems that are designed to be easy to use by older adults will also be easy to use by other user groups" (Fisk et al., 2009, p. 244). There are, however, some elements of the overall design of IVR applications in which performance of older and younger adults differs. For example, in a study of navigational support aids designed to reduce memory demands, older participants benefited more from a printed graphical aid, while younger participants performed better with a screen phone—a phone that included a small display (Sharit, Czaja, Nair, & Lee, 2003). For IVRs intended primarily for use by older adults it is reasonable to accommodate their needs with regard to pacing of speech and pauses.

Although there are differences in the hearing of young and older adults, this does not affect design. Normal young adults can hear pure tones with frequencies up to 15 kHz (kiloHertz, or thousands of vibrations per second). Age does not usually affect the ability to hear low-frequency tones, but many older adults cannot hear tones with frequencies higher than 4 kHz. Fortunately, the typical frequency range of speech vowel sounds is from 0.1 to 4 kHz. However, some consonants, especially fricatives such as /s/ and /f/, can hit frequencies above 7 kHz (Denes & Pinson, 1993). This property of language isn't just a problem for older adults—standard phone lines have a bandwidth of only 3.2 kHz (Lewis, 2004). One consequence of this for SUI design is to avoid critical distinctions in speech input or output that depend on hearing high-frequency consonants.

Relative to younger adults, older adults have more difficulty with fine motor movements, requiring more time to complete the movement and having less precision. A rule of thumb for the estimation of movement times for older adults is to double the movement times of younger adults (Fisk et al., 2009). For the purpose of SUI design, this amplifies the value of speech over touchtone, especially when the keypad is in the handset.

Cognitive assessment consistently shows age-related decline in working (short-term) memory and attention, especially auditory selective attention (Barr & Giambra, 1990), but not so much for long-term semantic or procedural memory (Fisk et al., 2009). The typical strategy of SUI designers to minimize demand on working memory and attention, due to the reliance of IVR applications on the human auditory channel, is consistent with good design for older adults. To take full advantage of semantic and procedural memory for any user it is important to design in accordance with population stereotypes, matching the order of events and the conversational elements to user expectation.

Other individual differences hypothesized to affect human performance include:

- Introversion/extroversion
- Field dependence/independence
- Internal/external locus of control

- Imagery
- Spatial ability
- Type A/B personality
- Ambiguity tolerance
- Memory span
- Learning styles
- Novice/expert
- Gender

A recent review of research on learning styles concluded that there is little evidence that hypothesized learning styles (words versus pictures versus speech; analysis versus listening) have significant effects on learning (Pashler, McDaniel, Rohrer, & Bjork, 2009). Investigations of the effects of these individual differences on human–computer interaction, other than field dependence/independence, have rarely established crossed interactions with elements of design (Lewis, 2006). In other words, the general finding with regard to individual differences is that systems designed to accommodate poorer performers either have no effect on better performers or actually improve performance for all users. This is a fortunate outcome for designers, because to accurately identify individual users as being introverted or extroverted, having high or low memory span, or having internal or external locus of control requires the administration of specialized psychological tests, which is not practical. It is important to keep in mind, however, that there has been little research on the effects of these individual differences when using IVRs.

Nass and Brave (2005) reported a number of studies of the effect of individual differences when using speech applications. For the most part, the studies were replications of classic social psychology studies of interactions between humans, but replacing one of the humans with a speech-enabled computer—a variation of the "computers as social actors" paradigm. One of their key findings was the replication of the "similarity attraction" effect—that people are attracted to other people who are similar to themselves so, for example, extroverts would prefer an extroverted user interface and males would prefer to hear a male voice when they call an IVR. It turns out, however, that it is difficult to apply many of these findings to user interface design.

For example, at the start of a call an IVR usually has no information about the caller other than, for some systems, the caller's ANI (automatic number identification), which does not guarantee knowledge about the true caller. To continue the script of the call, the IVR must say something to the caller. If the caller is female and the IVR voice is female, then similarity attraction should theoretically enhance the caller experience. If the caller is male, then according to the principle of similarity attraction the caller experience will be something less than optimal.

You could argue that the majority should rule—that the choice of the gender of the voice should match the gender of the majority of the population of expected callers. This, however, runs counter to the population stereotype in the United States that operators are usually female. Also, it is not clear whether the findings of Nass and Brave (2005) generalize from the replication of social psychology experiments to having strong effects on systems in use, where it is possible that these social media variables have little influence on the qualities of an otherwise usable system—one that is efficient, effective, and pleasant (Balentine, 2007).

In 2001, I ran an experiment to compare revised Mean Opinion Scale ratings (MOS-R) (Lewis, 2001b) of four artificial voices, two male and two female. Sixteen participants (eight male, eight female) listened to and rated all four voices in counterbalanced order. Although it was not the purpose of the study to test the similarity attraction hypothesis, the data were available to see if similarity attraction significantly influenced the participants' ratings. If listeners tended to strongly prefer voices that match their gender, then an analysis of variance should have detected a significant Listener Gender by Voice Gender interaction, with males giving relatively better ratings to male voices and females giving relatively better ratings to female voices. This, however, did not happen—the Listener Gender by Voice Gender interaction was not significant ($F(1,8) = 1.4$, $p = 0.26$). Because the means for female voices included the best- and worst-rated voices, which might have obscured the effect, I did the analysis again, using only the data from the highest-rated female and male voices. Again, the Listener Gender by Voice Gender interaction was not significant ($F(1,8) = 0.4$, $p = 0.55$). Figure 6.2 shows the Listener Gender by Voice Gender interactions (both best and combined).

Couper, Singer, and Tourangeau (2004) studied the influence of male and female artificial voices on more than 1000 respondents to an IVR survey on sensitive topics. Measurement variables included respondents' reactions to the different voices and abandoned call rates. They found no statistically significant results related to the gender of the voices. In particular, there were no significant gender-of-voice by gender-of-respondent interactions. The results from a disclosure experiment reported by Nass, Robles, Heenan, Bienstock, and Treinen (2003) indicated no voice gender by caller gender interaction for ratings of comfort with disclosure or actual disclosure behavior. "Why such strong effects of humanizing cues are produced in laboratory studies but not in the field is an issue for further investigation. … Across these studies, little evidence is found to support the 'computers as social actors' thesis, at least insofar as it is operationalized in a survey setting" (Couper et al., 2004, p. 567).

What all this indicates for the SUI designer is that it is not necessary to gather an expensive, exhaustive profile of expected callers' individual differences. On the other hand, the individual difference of expertise, both

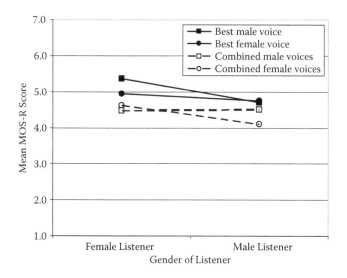

FIGURE 6.2
Male and female MOS-R ratings of male and female TTS voices.

familiarity with the task and frequency of usage of the system, clearly affects user performance and behavior (Novick, Hansen, Sutton, & Marshall, 1999). According to Mayer (1997), relative to novices, experts have better knowledge of syntax (the valid ways of communicating with the system), an integrated conceptual model of the system, more categories for more types of routines, and higher-level plans.

Fortunately, because its determination only requires usage data rather than sophisticated psychological testing, estimating the expertise of callers to an IVR is a relatively inexpensive exercise. Furthermore, in most cases SUIs designed for experts are also good for novices because both experts and novices benefit from efficiencies that result from concise initial prompting. Differences in typical caller expertise can, however, affect some SUI design decisions, such as whether to place examples before or after an open-ended prompt like "How may I help you?" (Sheeder & Balogh, 2003).

Based on the previous discussion of research related to user characteristics, some of the key user research questions to answer for IVR architecture and design are:

- How frequently will callers use the system?
- What is their motivation for using the system?
- In what type of environment(s) will callers typically use the system (quiet office, outdoors, noisy shopping mall)?
- What type of telephone connection will most callers have (land-line, cordless, cellular, IP telephone)?

- What language(s) will callers speak?
- How comfortable are typical callers with automated ("self-service") applications?
- To what extent will the calling population include older adults?

Analyzing User Tasks

One of the most important human factors activities in support of interaction design in general, including the design of IVR applications, is task analysis. To perform task analysis it is necessary to specify who will carry out the task, what the contents of the task are, and why the user must perform the task. The historical roots of task analysis lie in the time-and-motion studies of the early twentieth century (Roscoe, 1997). For decades, however, it has also included analysis of the cognitive aspects of tasks, such as identification of the points in task performance that threaten to overload human working memory capabilities. When determining what tasks the IVR application should support, consider:

- What are the most common tasks callers perform? What tasks are less common? Are callers familiar with the tasks they will need to complete?
- Will callers be able to perform these tasks by other means (for example, in person, using a visual Web interface, or by calling a customer service representative)?
- Will callers have the option of transferring to an agent?
- What words and phrases do callers typically use to describe the tasks and items covered in the IVR application?

Note the importance of maintaining a strong focus on the caller when analyzing tasks (Larson, 2005). Gardner-Bonneau (1999, p. 154) provided an example of what happens when designers fail to analyze tasks from the perspective of the caller:

> I once worked with an IVR dialogue for reporting power outage problems to an electric company. Clearly, the dialogue, as shown to me initially, was written from the point of view of a utility repair person. One of the prompts to the caller asked whether the line from the house to the utility pole was damaged. The only responses available to the caller were "yes" and "no." If I'm a caller reporting a power outage, presumably, my power is out. I don't have lights and it may be dark outside. How can I possibly be expected to answer this question as posed to me in the IVR dialogue? Obviously, the dialogue developer did not "stand in my shoes" when he or she put pen to paper.

One of the first design considerations for speech applications is the suitability of speech for the task. Consider using speech if:

- Users are motivated to use the speech application because it:
  - Saves them time or money
  - Is always available
  - Provides access to features not available through other means
  - Allows them to remain anonymous and to avoid discussing sensitive subjects with a human
- Users will not have access to a computer or prefer not to use the Internet
- Users want to use the application in a hands-busy or eyes-busy environment (Halstead-Nussloch, 1989), such as when the user is mobile (Novick et al., 1999)
- Users are visually impaired or have limited use of their hands

Avoid using speech if:

- The nature of the application requires graphics or other visuals such as maps, physical parts, or images, or of physical actions such as assembly instructions or dance movements (Novick et al., 1999)
- Users will operate the application in extremely noisy environments (for example, simultaneous conversations, background noise, etc.)
- Users are hearing impaired, have difficulty speaking, or are in an environment that prohibits speech (for example, a courtroom)

Larson's (2005) first of 10 guidelines for designing successful speech user interfaces is, "You can't design what you can't define." In addition to understanding customers and their needs, it is also important to know the business goals and rules that affect the target tasks. Polkosky (2010) cited spaghetti-like business logic as a common problem that makes it difficult if not impossible to craft a usable SUI. Thus one of the criteria for speech-enabling a task is that it has clear, noncontradictory business rules.

Kaiser, Krogh, Leathem, McTernan, Nelson, Parks, and Turney (2008) highlighted the importance of understanding the overall user experience—what happens before, during, and after the call to the IVR. Field research methods that can shed light on the user experience include observations of callers interacting with the IVR (either current or prototype), analysis of whole-call recordings (preferably capturing the system–caller interaction and, if routed to an agent, the agent–caller interaction), interviews or surveys with representative callers, and studies of channel preferences and transitions as a function of task (where "channels" refers to the various means customers have for communicating with an enterprise—phone, e-mail, Internet, etc.). The design of IVR tasks also benefits from an understanding of current enterprise branding, Web-based task automation, agent training materials, IVR and call center reports, end-to-end workflow processes and policies, and feedback from call center agents. To as great an extent as possible, SUI designers should "keep it simple," using backend information to limit the number of dialog steps presented to callers (Suhm, 2008).

Another aspect of task analysis to consider when speech-enabling a process is that partial automation of a process can provide significant benefits to an enterprise (Larson, 2005; Suhm, 2008). A common metric for IVR tasks is that of "containment"—the percentage of attempts at the task that a caller completely finishes in the application without any assistance from a call center agent. Too much emphasis on containment obscures the benefits, especially the time saved, by every element of a task completed in the IVR before the transfer to an agent, as long as that information gets transmitted to the agent upon transfer. For example, a reservation application on which I worked had about 15% containment for new reservations but, on average, provided five of the eighteen elements required to book the reservation, resulting in significant saving of agent time (a reduction of about 25%). Kaiser et al. (2008) reported similar savings (15%–20%) for partial automation of member calls to a health insurance call center after revising the business case based on the value of partial automation.

If there is no more direct way to get information about caller goals and behaviors, subject matter experts in the enterprise are always willing to provide their opinions, but designers must be very cautious about relying on these opinions unless they have an empirical basis. When I began working on the reservations application mentioned previously, subject matter experts from the enterprise provided a number of assumptions for the project. Table 6.1 shows the contrast between the subject matter experts' assumptions and the data gathered by reviewing 50 calls received by one of their call center agents.

As illustrated in the table and in Figure 6.3, there was a clear disconnect between some of the assumptions and the empirical findings. Gathering this information about the tasks early in the design process led to revision of the initial assumptions offered by the subject matter experts, which would have misled rather than informed the design.

A sample size of 50 calls is not sufficient to obtain precise measurements of percentages, but it is sufficient to get a quick sense of whether or not assumptions are reasonable (Lewis, 1996). For example, the lower limit of a 90% confidence interval (Sauro & Lewis, 2005) around the estimated percentage of new reservations for which the caller knew the membership number (18 of 19 cases, a point estimate of about 95%) was 78%. Thus even with a relatively small sample size, the evidence was overwhelmingly against the assumption that most callers would not have their membership numbers available when making a reservation.

TABLE 6.1

Assumptions and Findings about Caller Behavior When Making Car Reservations

Assumption	Finding
1: Callers would be members of their loyalty program.	*True*: The initial release was designed for members so we could take advantage of the information in their user profiles—all others transferred to an agent.
2: Most members have a very good idea about the rental they intend to make before placing a call and provide almost all of the information required for a rental (about 18 chunks, or tokens, of information) in their initial statement to the agent (for example, "My name is Joe Smith, and I need to get a car from the Miami airport on July second, about 6:00 p.m., returning on July sixth at about noon."). This example covers nine tokens of information—make reservation, need to rent a car, renter name, pickup location, pickup date, pickup time, return date, return time, implied return to pickup location—about half of the reservation in one continuous statement.	*False*: Members typically had all of the information they needed to make a reservation, but in most cases indicated in their initial statement to the agent (1) the purpose of the call (make, change, cancel) and (2) 75% of the time provided 0 or 1 additional token of information, most often, the desired rental location. See Figure 6.3 for a graph of the distribution of tokens as a function of new vs. existing reservation task.
3: Members have online profiles of preferred reservation, and will usually want their rentals to conform to the preferences in their profiles.	*True*: All callers in this sample wanted new reservations that conformed to their profile.
4: Most members making new reservations would not have their membership numbers available.	*False*: Members making new reservations had their membership number available 95% of the time.
5: Most members changing or canceling existing reservations would not have their reservation numbers available.	*False*: Members changing existing reservations had their reservation numbers available 73% of the time.
6: The best call flow would begin by getting the membership number, with a strategy of transferring to an agent if the member did not have the membership number available.	*False*: The best call flow should delay requesting the membership number until after collecting all profile-independent tokens to maximize the effectiveness of partial automation of the process.

Design Phase

The goal of this phase is to work out the details of the SUI design. This involves:

- Developing the conceptual design
- Making high-level decisions
- Making low-level decisions
- Defining the high-level call flow
- Creating the detailed dialog specification
- Prototyping

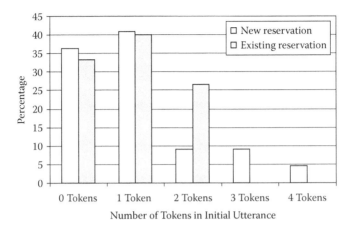

FIGURE 6.3
Number of tokens in the initial utterance of 50 car-rental calls.

Developing the Conceptual Design

A common approach to conceptual design is to write high-level scripts of proposed user–system interactions called vision clips. The focus is on the user–system interaction in key parts of important tasks. This design activity, also called "example dialogs," is an inexpensive first step to take prior to any large investment of resources (Sinha, Klemmer, & Landay, 2002).

In a vision clip, designers create preliminary samples of conversations to promote discussions between the designer and the customer regarding the customer's task and user interface expectations. Ideally, record these scripts so stakeholders can review the sound and feel of the proposed interactions. Designers of vision clips should be very familiar with the capabilities of the speech technologies to avoid preparing vision clips that would be difficult or impossible to deploy as applications. Table 6.2 illustrates the scripting for a vision clip.

Although it may increase the expense of the vision clips a bit, it can be helpful to have the designated voice talent record the audio segments to use for the system voice in the vision clips. If the voice talent has not yet been selected, then this can serve as an impetus for making the selection. Voice talent agencies are usually highly motivated to work with SUI designers and clients to help them select a voice that has good properties for use in speech applications and is consistent with enterprise branding (Graham, 2005, 2010).

It usually works well to provide clients with samples of three or four voices so they can choose the voice they feel best represents their company. It is also a good idea to make sure that the person who is the final decision maker for this aspect of the IVR participates in choosing the voice. Trust me—you do not want to be in a meeting where you're presenting the working version of the application (including all professional recordings) to the senior vice

TABLE 6.2

Example of Vision Clip Scripting: Find a Store

System/Caller	Scenario: Caller wants to find a store, is satisfied with first one
System:	Welcome to the Electric Razor Customer Care Line. <AudioLogo based on branding tone> Calls may be monitored and recorded. I can help you find a store, get replacement head info, or make a purchase. What would you like to do?
Caller:	Find a store.
System:	OK, Store Locator. What's your ZIP code?
Caller:	3 3 4 8 7.
System:	The closest store is: Fred's Razors. Located at: 1111 Federal Highway, Boca Raton, Florida. Their telephone number is: 561-555-1234. You can call them for business hours and local driving directions. Please select Repeat, Next Location, Change Zip Code, Start Over, or Exit.
Caller:	Exit.
System:	Do you want to end this call?
Caller:	Yes.
System	Thanks for calling the Electric Razor Customer Care Line. You can find us on the Web at eri.com. Goodbye!

president in charge of customer care who, upon hearing the voice for the first time, says, "I hate it. We need a different voice."

Making High-Level Decisions

High-level application decisions include selecting an appropriate user interface, a barge-in style, using recorded or synthesized speech, audio formatting, type of grammars, prompting style, use of touchtone, global navigation commands, access to human agents, and help style. See Chapter 7 for details.

Making Low-Level Decisions

After making the high-level decisions, the lower-level system decisions address issues such as design of the introduction, timing issues, general dialog design, design of menus and prompts, specific grammars, and confirming user input. Chapter 8 provides information and examples to help with these decisions.

Defining the High-Level Call Flow

The next step is to work out a high-level depiction of the call flow that maps the interaction between the application and the caller. For example:

- From the task analysis, what are the key tasks and task steps that the caller must perform?

- When the caller answers (or fails to answer), what should the application do?
- What will the application do automatically?

The call flow should account for typical responses, unusual responses, and any error conditions that might occur. It isn't necessary to include specific prompt wording at this level of design. Rather, to avoid duplication of the specific prompt and message information that will be part of the detailed dialog specification, which can make it difficult to keep the documents in sync when making changes, the high-level call flow should be a graphical representation of the gist of the flow. For an example of a high-level call flow, see the Sample Design Documents at the end of this chapter.

Creating the Detailed Dialog Specification

After defining the high-level call flow, it is time to draft a detailed dialog specification (DDS). The specification should include all of the text that will be spoken by the application, as well as expected user responses. For an example of a DDS, see the Sample Design Documents at the end of this chapter.

As a check on the completeness of the DDS, it's a good practice to construct scenarios of use. In planning the application, you should have created a list of the tasks that the application can perform. Use this list to create scenarios of use, where a scenario of use is a concrete task to accomplish with the application, such as "Transfer $100 from savings to checking." Use your script to walk through the interaction and to create the expected call flow for the scenario (similar to the process of creating vision clips). This exercise can be very helpful in early identification of scripting problems. For examples of scenarios of use derived from a DDS, see the Sample Design Documents at the end of this chapter.

Prototyping

The goal of this phase is to create a low-fidelity prototype of the application, leaving the design flexible enough to accommodate changes in prompts and dialog flow in response to problems discovered during building and testing the prototype.

For the first iteration, it is common to use a technique known as "Wizard of Oz" testing (Edlund, Gustafson, Heldner, & Hjalmarsson, 2008; Hajdinjak & Mihelič, 2004; Kelley, 1985). At its simplest, this technique requires no programming—just a paper script, a set of planned tasks for the caller to complete, equipment to record the session, and two people: one to play the role of the user and the other (the "wizard") to play the role of the IVR. Here's how it works:

- The script should include the proposed introduction, prompts, list of global commands, and all planned help messages.

- The two participants should be physically separated so they cannot use visual cues to communicate. Ideally, they speak to one another by telephone. In a pinch, they can sit back to back in the same room.

- The wizard must be very familiar with the script. The user should never see the script.

- The user telephones (or pretends to telephone) the wizard, who begins reading the script aloud. The user responds to the prompts, and the wizard continues the dialog based on the scripted response to the user's utterance.

"Wizard of Oz" testing helps with the identification of usability problems in the script and task flow before committing any of the design to code (Sadowski & Lewis, 2001). The results can also guide grammar development. However, it cannot detect other types of usability problems, such as recognition, audio quality, and turntaking or other timing problems. Addressing these types of problems requires a working prototype (Sadowski, 2001).

The introduction of VoiceXML (Sharma & Kunins, 2002) substantially reduced the effort required to produce a working prototype suitable for demonstrating proofs of concept and conducting usability tests (Lewis, 2006, 2008). Chapter 9 provides an extended example of building a VoiceXML prototype for these purposes, with tips on prototyping in accordance with the SUI design principles discussed in Chapters 7 and 8.

Development Phase

The findings from "Wizard of Oz" and usability tests of prototypes conducted during the design phase feed back into the final versions of the high-level call flow and DDS. The primary activity during the development phase is the implementation of the DDS in hardware and software. The paper design becomes a real application, with fully specified grammars or language models, interfaces to backend databases, Web services, and other software systems, coded and unit-tested. Much of the work in this phase is done by software developers.

During this phase, it often falls upon the SUI designer to:

- Compile the recording manifest, creating a list of all of the segments submitted to the voice talent agency at least a day before the scheduled recording session (see the recording manifest example in the Sample Design Documents at the end of this chapter)

- Make arrangements with the talent agency for the recording session, including the time and day with the designated voice talent
- Coach the voice talent during the recording session, fostering a relaxed environment and listening attentively throughout the session
- Review and, if necessary, edit the recorded audio before deployment
- Identify a few key phrases to use for future sessions to help the voice talent recapture the desired style and tone, and to ensure consistency between recording sessions

Test Phase

As in the development phase, much of the work in the test phase is done by software developers, who install the new application on test servers and then conduct functional verification tests (FVT), system verification tests (SVT), stress (load) tests, and long-run tests. During this phase and depending on their professional backgrounds, SUI designers participate in SVT, customer acceptance tests, grammar tests, and usability tests.

System Verification and Customer Acceptance Tests

The purpose of SVT is to ensure that the system works as designed. Based on previous task analyses, SUI designers can help with the construction of the test cases used to verify the system operation. It can be valuable for SUI designers to participate as SVT testers because they can detect problems that might otherwise go undetected, such as problems in the connection points of audio segments, incorrect prosody in audio segments, and timing issues.

Strictly speaking, clients should independently conduct their customer acceptance tests (similar to SVT, conducted for the purpose of ensuring conformance to the specified design). In practice, they often rely upon the design team, including the SUI designers, to assist in the development and execution of the test. At the time of the customer acceptance test, modifications done in response to problems discovered during SVT should be in place, so this provides another opportunity for SUI designers to ensure that the system is working as designed.

Grammar Testing

SUI designers who have a background in experimental psychology have the skills to plan and conduct user tests of grammars and/or statistical language models. These tests can begin as soon as the grammar is available, and initial tests should take place well before the application is running. Grammar testing of SUIs should address both habitability (coverage) and recognition accuracy. Habitability is the technical term for the ease, naturalness, and effectiveness with which users can use language when working with an NL system (Watt, 1968). The purpose of habitability testing is to measure the extent to which the grammar accepts the utterances that users are likely to make when speaking to the system. The purpose of recognition accuracy testing is to measure how well the system can interpret utterances that it should recognize. Both habitability and recognition accuracy are important components of system usability for spoken NL user interfaces.

After analyzing various approaches to testing habitability and recognition accuracy, it appears that the best order for testing these system properties is habitability first, followed by recognition accuracy testing. The reason for this is that the process of testing habitability has as an output a set of user-generated phrases. Habitability testing reveals which of these phrases the system is capable of recognizing and which are not legal utterances for the system. Developers can use this information to broaden the language model of the system, increasing its habitability, or to narrow the language model, reducing the odds of misrecognition. Once the developers have extended the application's flexibility, the final habitability score is the percentage of user-generated phrases that are legal expressions in the application.

After completing the development activities that affect the system's language model, the set of legal user-generated phrases becomes an excellent set for testing the system's recognition accuracy. However, there is no guarantee that the set of user-generated phrases effectively exercises the boundaries of the system's capabilities. To address this, it is necessary to develop a complementary set of test phrases based on analysis of the system's specified capabilities. There are a number of ways to approach this development, with the best way depending on the available system development tools. Testing with these two complementary sets (user generated for a set of natural test phrases and developer generated for a set of comprehensive test phrases) should provide reasonable coverage for an efficient assessment of an application's grammars.

There are four domains within which users must stay when speaking to an NL application:

- Conceptual (you can't ask a traffic report system about stock quotes)
- Functional (the system might handle requests that contain multiple tokens, or might require users to create single-token phrases)
- Syntactic (the system might allow many or few paraphrases for commands)
- Lexical (the system might allow many or few synonyms for the words in the application)

"A natural language system must be made habitable in all four domains because it will be difficult for users to learn which domain is violated when the system rejects an expression" (Ogden & Bernick, 1997, p. 139). For example, consider system rejection of the phrase, "Provide today's arrival times and destinations for Delta Flight 134." If the user was talking to a stock quote application, then the violation was in the conceptual domain. If the system could provide arrival times or destinations, but not both from a single request, then the violation was in the functional domain. If the system would have understood the request in a different syntactic form ("What are the arrival times and destinations for Delta Flight 134 today?"), then the violation was in the syntactic domain. Finally, if the system would have understood the command if the user had substituted "list" for "provide," then the violation was in the lexical domain. Ogden and Bernick (1997) describe several habitability evaluation issues, including user selection, training, task generation, and task presentation.

As with any user evaluation, the participants should be representative of the intended user population. Participants should receive training in the use of the system that is representative of the training that will be available to the intended user population. Tasks can be either user or tester generated, but tester-generated tasks based on the functional specification have greater assurance of more comprehensive functional coverage than user-generated tasks. On the other hand, tester-generated tasks will not address conceptual habitability because the tester typically knows the conceptual limitations of the system. If possible, the tester should present tasks to the user with tables or graphs, asking the participant to say what he or she would say to the system to get the information needed to put information into the cells of the table or to complete the task indicated on the graph. Figure 6.4 shows some examples of these types of task representations.

The advantage of this type of task representation is that it induces less linguistic bias to participants than writing the task in words (although it is sometimes impossible to construct these types of scenarios without any words). If you must use text to set up the test case strive for simple language. (For example, "You need to report a problem. You ordered Juno using your on-demand movie service. While you were watching, it stopped playing.") If possible construct different versions of the test case using different wording.

The more people participating in the prototype study, the better, up to the point of diminishing returns where additional participants provide few additional phrases. There are no published experiments on the effect of running different numbers of participants on the breadth of expressions produced, but similar work in problem–discovery usability studies (Lewis, 1994) suggests that six to ten participants would be a reasonable sample size for planning purposes. If the application's audience includes very young or old users, then the sample should include participants whose ages span the relevant range. This also applies to applications intended for use by people

TABULAR – AIRLINE FLIGHTS

Carrier	Flight No.	Departure City	Departure Time	Arrival City	Arrival Time
Delta	123	Miami	12:00 PM	Atlanta	
	876	Dallas	8:00 AM	Seattle	11:00 AM

Potential user input

What time does Delta flight 123 arrive in Atlanta?
Tell me the airlines that leave Dallas for Seattle around 8:00 AM?

GRAPHICAL – CURRENCY CONVERSION

| $100 US | = | ? Euros |

Potential user input

How many Euros can you get for one hundred US dollars?
Please tell me the current rate for converting US dollars to Euros.

FIGURE 6.4
Examples of tabular and graphic task representation.

speaking different dialects, or who have other relevant cultural or experiential differences. The more of these types of user characteristics that it is important to sample, the larger the sample size will need to be. Testers need to balance the need for adequate coverage of user characteristics with the available resources.

After gathering the participants' utterances obtained during habitability testing, the next step is to determine which utterances the application can recognize. The precise way to do this depends on the development tools available for the system. For systems with formal grammars, the toolkits usually provide a tool for checking to see if a phrase is in grammar. Other NL toolkits could provide tools for checking to see that all the words are in the language model, or could even check to see that the system parses the utterances and correctly extracts the tokens. Developers can use the output of this analysis in two ways:

- Alter the system so it can correctly interpret more of the utterances
- Use the correctly interpreted utterances as a phrase set for testing recognition accuracy (described in more detail in the following section)

Three questions to answer before conducting a recognition accuracy test are:

- How many phrases should you collect from each speaker?
- How do you select the phrases?
- How many speakers should participate in the test?

TABLE 6.3

Strategies for Developing the Set of Test Phrases for Recognition Accuracy Testing

Strategy	Description
1. Use application functions to guide test set development	It is possible to use a grammar to produce the smallest possible set of test cases that can comprehensively test the grammar with regard to verifying correct output to a translation rule interpreter or other parsing system (in other words, based on functions). This minimal test set is also appropriate for testing accuracy because it generates the full range of the types of phrases that are legal in the grammar and uses all of the words in the grammar. See Lewis (2000) for a method of constructing this minimal test set.
2. Focus on frequently selected items from large lists	If a grammar includes very large lists (such as stocks listed in stock exchanges), then it is necessary to limit the phrases using the members of these lists to a manageable number. For example, there are roughly 8000 airports in the world, but almost all of the world's airport traffic moves through about 600 of those airports. It seems reasonable to limit accuracy testing to the 600 airports that account for more than 99% of the world's airport traffic.
3. Use the output of habitability testing	This approach has the advantage of providing reasonable assurance that the test phrases are representative of the types of things users are likely to say when using the application—something that no other strategy can provide. The disadvantage is that there is no guarantee that collecting phrases from a relatively small group of participants will provide the same extent of coverage as the minimal test set of Strategy 1.
4. Random generation of test set phrases	Some toolkits provide functions for producing randomly generated phrases from the grammar. This is an easy method, but there is no guarantee that the phrases generated will provide an efficient evaluation of potential acoustic confusability, especially among the members of item and synonym lists. Also, the test set might not provide adequate functional, syntactic, or lexical coverage.
5. Use phrases from the natural language application's training set	It might be tempting to draw the test phrases from the materials used to generate (train) the natural language application. Using test phrases drawn from the source material, though, would very likely lead to an overestimation of the accuracy of the system due to the unrealistically close match between the words in the phrases and the content of the language model (or finite-state grammar) produced by the natural application development system. If there is a set of held-out phrases— phrases that could have been part of the training set but were held out for the purpose of evaluating the language model—then it is legitimate to continue using those phrases as a source for recognition testing.

The first two questions apply to the development of the set of test phrases, with various strategies for this development, listed in Table 6.3. The answer to the third question depends on the standard statistical concepts of measurement variability and required measurement precision (Lewis, 2006). All things being equal, larger sample sizes are better than smaller ones. It is, however, better to run small samples than no samples at all. Our evaluations of recognition experiments conducted over several years have indicated that eight speakers will typically yield measurements with acceptable precision, although we have successfully used as few as four speakers.

Regardless of which strategy a tester uses to develop the set of test phrases, to ensure adequate precision of measurement the test should:

- Include at least six to twelve speakers with equal (or as equal as possible) numbers of male and female speakers (and addressing other coverage issues as required).
- Have at least 100 different phrases in the test set.

The disadvantage of a Strategy 3 test set is that it might not examine the full range of functions, syntaxes, and words that are legal in the application. Therefore, if it is possible, the tester should also develop a Strategy 1 test set, and use both test sets to assess the application's recognition accuracy. This is the best approach because the two types of test sets have complementary strengths and weaknesses.

If the application includes the use of large lists (for example, stock names), then the tester should develop test sets for each large list, testing only the most important members of the list to determine if any pairs of important members have high acoustic confusability and if the system confuses any less important members with important members.

This approach, using multiple test sets with different and complementary evaluative properties, should provide the test coverage necessary to obtain a comprehensive assessment of both recognition accuracy and habitability for natural command and natural language applications.

Dealing with Recognition Problems

Some phrases will have better recognition than other phrases. Analysis of the phrases with relatively high-failure rates across participants can lead to changes to improve system accuracy. These changes can include:

- Modifying the grammar/language model of the system
- Adding additional pronunciations for frequently misrecognized words in the system

Note any consistent recognition problems. The most common cause of recognition problems is acoustic confusability among the currently active phrases (Fosler-Lussier, Amdal, & Kuo, 2005). For example, both Madison and Addison are names of U.S. airports. Thus, due to coarticulation, these potential user inputs to a travel application are highly confusable:

User: Flying from Madison
User: Flying from Addison

Sometimes there is nothing you can do when this happens. Other times you can correct the problem by:

- Using a synonym for one of the terms, changing grammars, prompts, and help messages as required. For example, if the system is confusing "no" and "new," you might be able to replace "new" with "recent," depending on the application's context (with corresponding changes in prompts to guide successful interaction).

- Adding a word to one or more of the choices, changing grammars, prompts, and help messages as required. For the Madison/Addison airport confusion, you could make states optional in the grammar for most cities, but require the state for low-traffic airports that have acoustic confusability with higher-traffic airports.

- Plan for disambiguation by writing code that includes or accesses data about typical acoustic confusions. For example:

System: Flying from?

User: Los Angeles <not flagged as confusable>

System: Flying to?

User: Newark <flagged as confusable with New York>

System: Newark, New Jersey, or New York, New York?

User: Newark, New Jersey

Usability Testing

Habitability and recognition testing provide important information, but are not sufficient to fully address the usability of performing tasks with the application. For software development in general, one of the most important test and evaluation innovations of the past 30 years is usability testing (Lewis, 2006). In reference to software the term "usability" first appeared in 1979, and came into general use in the early 1980s, displacing the earlier related terms of "user friendliness" and "ease of use." Recent surveys of usability practitioners indicate that it is one of their most frequent activities in commercial software development, second only to iterative design, with which it has a close relationship (Vredenburg, Mao, Smith, & Carey, 2002).

Usability is not a property of a person or thing, so there is no instrument that gives any absolute measurement of product usability. Usability is an emergent property dependent on the interactions among users, products, tasks and environments. Adding to confusion regarding the term "usability testing" practitioners use the term "usability" in two related but distinct ways. For "summative" (measurement based) usability testing, the primary focus of usability is on measurements related to the accomplishment of global task goals. For "formative" (diagnostic) usability testing, the focus is on the detection and elimination of usability problems.

During a usability test, one or more observers watch one or more participants perform specified tasks with the product (pre-release or final version) in a specified test environment. Thus usability testing is different from other user-centered design methods (Dumas & Salzman, 2006). Interviews, either individual or in focus groups, do not include the performance of work-like tasks. Expert evaluations, heuristic evaluations, and other usability inspection methods do not include the observation of users or potential users performing work-like tasks, which is also true of techniques such as surveys and card sorting. Field studies typically involve the observation of users performing work-related tasks in target environments, but observers have little or no control over the target tasks and environments (which is not necessarily a bad thing, but is a defining difference).

Although there are many potential measurements associated with usability tests, in practice, as recommended in the Common Industry Format for Usability Test Reports (ANSI, 2001; ISO, 2006) and supported by statistical analysis of multiple sets of usability test data (Sauro & Lewis, 2009), the fundamental global measurements for usability tasks (and by extension, for IVR transactions—Suhm, 2008) are:

- Successful task completion rates for a measure of effectiveness
- Mean task completion times for a measure of efficiency
- Mean participant satisfaction ratings, collected either on a task-by-task basis, at the end of a test session, or both.

Based on these three overarching goals, Bloom, Gilbert, Houwing, Hura, Issar, Kaiser, et al. (2005) identified ten key criteria for measuring effective voice user interfaces, summarized in Table 6.4.

The goals for summative usability tests should have an objective basis and shared acceptance across the various stakeholders. Data from pilot tests or previous comparable usability studies provide a reasonable objective basis for measurement goals (similar types of participants completing the same tasks under the same conditions). If previous data are not available an alternative is to define a set of shared goals based on general experience and negotiation with other stakeholders.

After establishing goals, the next step is to collect the data with representative participants performing the target tasks in the specified environment. Even when the focus of the test is on task-level measurement, test observers should record the details about the observed usability problems. It is better to run one usability test than not to run any at all. On the other hand the real power of usability testing derives from an iterative product development process. Ideally usability testing should begin early and occur repeatedly throughout the development cycle.

After test completion, the next step is data analysis and reporting. Summative studies typically report the standard task-level measurements such as completion

TABLE 6.4

Ten Key SUI Effectiveness Criteria

Criterion	Description
1. Caller satisfaction	Assessed at a minimum with one or two five-point Likert items, such as, "I was satisfied with the automated portion of this call" and "I was satisfied with the agent during this call." For more detailed evaluation of caller satisfaction use psychometrically qualified instruments such as the 34-item Subjective Assessment of Speech System Interfaces (SASSI) (Hone & Graham, 2000), the 11-item Pragmatic Rating Scale for Dialogues (Polkosky, 2002), or the 25-item Framework of SUI Service Quality (Polkosky, 2008).
2. Perceived ease of use	Assessed at a minimum with one five-point Likert item, such as "The application was easy to use." For more detailed evaluation of perceived ease of use, see the items for the User Goal Orientation and Customer Service Behavior factors of the Framework of SUI Service Quality (Polkosky, 2008).
3. Perceived quality of output	Assessed at a minimum with two five-point Likert items ("The voice was understandable" and "The voice sounded good.") For more detailed evaluation of voice quality, see the five items for the Speech Characteristics factor of the Framework of SUI Service Quality (Polkosky, 2008) or, for a multidimensional assessment, the 15-item MOS-X (Polkosky & Lewis, 2003).
4. Perceived first-call resolution rate	Assessed at a minimum with a yes or no answer to the question, "Did you accomplish your goal?"
5. Time-to-task	The time that callers spend in the IVR before they can begin the desired task. Lower values of time-to-task lead to greater caller satisfaction. Items that increase time-to-task include lengthy up-front instructions, references to a Web site, and marketing messages.
6. Task completion rate	The rate at which callers actually accomplish tasks (an objective measure in contrast to the subjective measure of perceived first-call resolution rate).
7. Task completion time	The time required for callers to complete tasks. Generally, shorter task times are better for both the caller and for the service provider.
8. Correct transfer rate	Percentage of transferred calls getting to the right agent.
9. Abandonment rate	Percentage of callers who hang up before completing a task. Ideally the design of the IVR and associated logging discriminates between expected (probably not a problem) and unexpected (probably a problem) disconnections.
10. Containment rate	Percentage of calls not transferred to human agents. Although this is a common metric, it is deeply flawed. See the discussion below for details (in the section on call log analysis in the tuning phase).

rates, completion times, and satisfaction, and will usually also include a list of the observed problems. Formative reports normally provide only a list of problems but can provide preliminary estimates of task-level measurements.

Because usability tests can reveal more problems than available resources can address, it is important to prioritize problems. A data-driven approach to prioritization takes into account problem-level data such as frequency of

TABLE 6.5

Sample Size Requirements for Problem Discovery (Formative) Studies

Problem Occurrence Probability	Cumulative Likelihood of Detecting the Problem at Least Once					
	0.50	0.75	0.85	0.90	0.95	0.99
0.01	68	136	186	225	289	418
0.05	14	27	37	44	57	82
0.10	7	13	18	22	28	40
0.15	5	9	12	14	18	26
0.25	3	5	7	8	11	15
0.50	1	2	3	4	5	7
0.90	1	1	1	1	2	2

occurrence, impact on the participant, and likelihood of usage. A common impact assessment method assigns scores depending on whether the problem (1) prevents task completion, (2) causes a significant delay or frustration, (3) has a relatively minor effect on task performance, or (4) is a suggestion. The combination of the frequency and impact data can produce an overall severity measurement for each problem. The most common use of quantitative measurements is the computation of means, standard deviations, and, ideally, confidence intervals for the comparison of observed to target measurements when targets are available. When targets are not available, the results can still be informative, for example, to aid in the estimation of future target measurements. For an example of a usability evaluation of a speech recognition IVR, see Lewis (2008).

When the cost of a test sample is low, there is little harm in oversampling. In usability testing, however, the cost of a sample (the cost of running an additional participant through the tasks) is relatively high, so sample size estimation is an important usability test activity. For task-level measurements, sample-size estimation is relatively straightforward using standard statistical techniques that require an estimate of the variance of the dependent measure(s) of interest and an idea of how precise the measurement must be (Lewis, 2006).

The past 15 years have seen a growing understanding of appropriate sample-size estimation methods for problem-discovery (formative) testing (Lewis, 1994, 2001a, 2006). Based on this research, Table 6.5 provides guidance for (1) estimating sample size needs before a formative usability test or (2) assessing its utility after finishing a formative usability test. Using the table to estimate a sample size requires a balance among problem occurrence probability, the cumulative likelihood of detecting the problem at least once, and the available resources. Smaller target problem occurrence probabilities require larger sample sizes. For example, to have a 50% chance of seeing problems that will occur only 1% of the time requires a sample size of 68 participants. In contrast, greater problem discovery likelihoods require larger sample sizes. For example, to have a 99% chance of seeing problems that will occur 25% of the time requires a sample size of 15 participants.

Use of the table can aid in the achievement of reasonable test plans. For example, it is clearly unrealistic to design a usability test with the goal of having a 99% chance of seeing problems that will occur only 1% of the time, because that would require testing 418 participants. A goal of having a 90% chance of seeing problems that will occur 25% of the time, which requires a sample size of eight participants, is more reasonable. Another consideration is whether there will be only one usability study or whether there will be a series of usability tests as part of iterative development.

Another use of the table is to gain an understanding of what a usability study was capable of detecting given a specific sample size. For example, suppose a practitioner has limited resources and can only run six participants through a usability study. Such a study is worth running because:

- The study will almost certainly detect problems that have a 90% likelihood of occurrence (the cumulative likelihood is 99% with only two participants).

- There is a high probability (between 95% and 99%) of detecting problems that have a 50% likelihood of occurrence.

- There is a reasonable chance (about 80% likely) of detecting problems that have a 25% likelihood of occurrence.

- The odds are about even for detecting problems with a 15% likelihood of occurrence (the required sample size at 50% is 5).

- The odds are a little less than even for detecting problems with a 10% likelihood of occurrence (the required sample size at 50% is 7).

- Although the odds are low that the study will detect a large number of the problems that have a likelihood of occurrence of 5% or 1% (the required sample sizes at 50% are 14 and 68, respectively), it will probably detect some of them.

Automated SUI Usability Testing Methods

There has been some interesting research in automated testing of speech applications. One line of research is in quality prediction models such as the PARadigm for DIalogue System Prediction (PARADISE). Another is in the development of simulated users.

PARADISE is a general decision-theoretic framework for evaluating spoken dialog systems (Walker, Kamm, & Litman, 2000; Walker, Litman, Kamm, & Abella, 1997, 1998). The goal of PARADISE is to allow comparison of different dialog strategies for accomplishing tasks with a speech application. The basic PARADISE method (Walker et al., 1997, 1998) is to create a model of performance as a weighted function of task success measures, with weights computed by correlating user satisfaction with performance. The method decouples what the speech application (automated agent) needs to know from how the agent does the task, enhancing the generalizability of the resulting function. The resulting PARADISE performance measure is a function that relates task success and dialogue costs (efficiency and qualitative measures) to user satisfaction. A series of experiments with different spoken dialog systems (Walker et al., 2000) showed that predictive PARADISE models created for one speech application explained significant portions of the variability of the response function in two other very different speech applications (55% and 36%, respectively, corresponding to correlations of 0.74 and 0.60). Subsequent independent attempts to apply the PARADISE model have highlighted the need for a better understanding of the correlation structure of the underlying measures and for a more reliable user-satisfaction measure (Hajdinjak & Mihelič, 2006).

Möller and his colleagues (Möller, Engelbrecht, & Schleicher, 2008; Möller & Skowronek, 2004; Möller, Smeele, Boland, & Krebber, 2007) have conducted research with PARADISE-like quality prediction models. In response to the criticism that the original PARADISE experiments used an ad hoc (rather than a psychometrically qualified) questionnaire to measure user satisfaction (Hone & Graham, 2000), Möller et al. (2007) used a modified version of the Subjective Assessment of Speech System Interfaces (SASSI). They concluded that the metrics they collected provided helpful information for system design and optimization, but did not lead to models that could accurately predict system usability or acceptability. Möller et al. (2008) compared different methods to predict quality and usability of spoken dialog systems. They found that linear regression models (such as PARADISE) and classification trees accounted for about 50% of the variance in their training data, and neural networks did somewhat better. Attempts to apply the resulting models to data obtained with different systems or user groups led to significant decline in predictive accuracy.

Another approach to automated usability evaluation of SUI applications is the development of simulated users (Engelbrecht, Quade, & Möller, 2009; López-Cózar, Callejas, & McTear, 2006; López-Cózar, De la Torre, Segura, & Rubio, 2003; López-Cózar, De la Torre, Segura, Rubio, & Sánchez, 2002). The simulated user of López-Cózar and colleagues includes transaction memory, a dialog manager (tied to a scenario corpus), and a response generator (tied to the scenario corpus and an utterance corpus). The scenario corpus contains a wide range of scenarios designed to be representative of user goals. The utterance corpus contains all of the utterances a user would need to interact with the speech application. Ideally, the utterances should include recordings from several different speakers and should represent the range of user utterances in the domain. López-Cózar et al. (2002) reported that even with the limitations of their first-generation simulated user, they were able to use the method to detect a problem in the confirmation strategy of a speech application. López-Cózar et al. (2006) extended the concept of a simulated user to different types of simulated users: very cooperative, cooperative, and not very cooperative, and reported some success in using the simulated users to identify problems in a speech application with lengthy and repetitive confirmation sequences for the correction of postal codes and addresses.

Although the PARADISE quality prediction and simulated user concepts are interesting, they are still active areas of research in the evaluation of spoken dialog systems and are not yet part of practical SUI design.

Deployment Phase

Software developers are responsible for much of the deployment phase. Their activities include:

- Setting up the environment with appropriate machines and software
- Training the enterprise with regard to the tasks its employees will perform with the application, such as generation and interpretation of system performance reports
- Configuring the application platform and tuning recognition grammars and engines

After completing all other testing and making the indicated changes to the application, but before full deployment, it is helpful to conduct a pilot test (Cohen et al., 2004). In pilot testing, a limited percentage of callers experience the new application. This test provides the first data of real callers using the new application to achieve real tasks. All the usage data (and possibly, post-call interviews with selected callers) provide information used to make the final changes to the application. SUI designers typically participate in the design of changes that affect the speech user interface. After making all the final changes, the software engineers integrate the application at the customer site, and all callers begin reaching the new, fully deployed application.

Tuning Phase

The tuning phase covers all post-deployment activities, such as adjustments to the application based on surveys, call log analysis, and end-to-end call monitoring. Application adjustments that are of interest to SUI designers include changes to the call flow, grammars, and the prompts and messages (Guinn, 2010). These changes can result from problem tracking reports of program bugs and design change requests for the current application, or can drive requirements for a future version of the application.

Surveys

After deploying an application, it is a common practice to survey callers by asking at the end of the call if they are interested in providing their opinion about the application. Most surveys include a few questions about the perceived quality of the application and an open-ended question for which the caller responses are recorded for transcription and analysis. When interpreting the results of this type of survey, it is important to keep in mind that the callers have selected themselves, which means that their responses are not necessarily representative of the population of all callers. Even worse, callers who experience significant difficulty with the system and disconnect never hear the invitation to take the survey and thus never have the opportunity to influence the measurements or to provide verbal feedback (Leppik & Leppik, 2005). To get a more representative sample, it is necessary to use a random selection process that includes callers who did not get to the end of the call.

Call Log Analysis

The data about the calls made to the system captured in call logs can be helpful in identifying weaknesses in the deployed application. Some common metrics, usually provided as global measures and broken down at each dialog step, are:

- Abandon rate
- Call containment
- First call resolution

The abandon rate is the percentage of callers who hang up on the application, typically tracked at each dialog step. It can be difficult to understand why a caller has disconnected the call, making a standard abandon rate difficult to interpret. For example, what does it mean if a caller hangs up during the application's introduction? Is that or is that not a problem that requires analysis and treatment? What if the application for a pay-as-you-go cellular provider starts by telling the caller how many minutes are left on the account, then the caller hangs up as the application plays its main menu? Does that mean that the caller's need was satisfied by the preliminary message, or does it mean that something in the main menu confused the caller? It can be helpful during detailed dialog design to define abandoned calls at each point in the dialog as expected (probably not a problem) or unexpected (probably a problem). Parsing the data in this way makes it easier for analysts reviewing reports to find problem areas more quickly.

Another common key performance indicator is call containment, also known as calls serviced in the IVR, self-service resolution, IVR utilization, IVR take-rate, or call retention. The typical definition of this metric is the percentage of calls resolved in the IVR without agent assistance. Although this is a common metric, it can be misleading because it ignores an important part of the purpose of an IVR—to route calls correctly—and it fails to account for benefits due to partial automation (Suhm, 2008). As Leppik (2006, p. 1) noted:

> There's a couple of unstated assumptions built into this metric which make it a poor way to measure IVR performance. The first assumption is that the system's only function is to automate customer calls. In truth, the IVR has a second, just as important (and maybe more important) function: to identify which customers' calls must go to an agent, and efficiently connect those customers to people who can help them. This latter group of calls is going to include the sales calls, billing errors, and already-upset customers who have been trying to resolve their problem for weeks. You can never automate those calls, and failing to identify them and get them off the IVR and into an agent's hands is as much of a system failure as sending a potential self-service call to a human.

The converse of call containment is the transfer rate—by definition, a contained call never transfers to an agent (Bloom et al., 2005). As with the abandon rate, it can be a difficult metric to interpret. For this reason, it can also be helpful to define for each point in the dialog whether a transfer is expected (probably not a problem) or unexpected (probably a problem), and to report these metrics separately.

Suhm (2008, p. 20) has warned analysts to be very careful when interpreting standard IVR reports:

> While often interpreted as the success rate for serving callers in an automated fashion, IVR take-rate is a poor measure of the effectiveness of an IVR, because callers hanging up in the IVR may not have received any useful information. In several large call centers we have seen that the majority of callers hanging up have actually received no useful information and therefore have not been served. For example, based on standard IVR reports, one call center believed that its IVR served more than 30% of the callers in the automated system. A detailed analysis based on end-to-end calls revealed that only 2% of all callers were actually served. Almost 20% hung up without receiving any useful information, and some 8% hung up while on hold for an agent.

For the call center overall (IVR and agent interaction), an important measurement that affects customer satisfaction is First Call Resolution—the ability of a call center to resolve a customer's request on the first call with no transfers and no callbacks. To measure First Call Resolution, it's necessary to have in place a centralized end-to-end logging and reporting system.

End-to-End Call Monitoring

There is simply no substitute for end-to-end call monitoring of randomly selected calls, hearing the interaction between the caller and the IVR (also called whole-call recordings). Random selection ensures that the sampled calls are representative of the entire population of calls. End-to-end calls to a deployed system are of tremendous value because they are real calls made by real customers who are trying to accomplish real tasks (Suhm, 2008; Suhm & Peterson, 2002). Data don't get any better than that. Even though the sample size of the data (the number of calls) in call logs will be substantially greater than the number of transcribed end-to-end calls, call logs cannot provide the richness of knowing exactly where in a prompt or message a caller barged in, or exactly where a poorly timed message caused the caller to stutter, resulting in a misrecognition.

Some premier systems can track and record calls end to end across the different components of the architecture, from the IVR introduction through transfers to one or more agents, to disconnection. Other systems do not record the interaction with the IVR, but do allow monitoring of the entire call, in which case it is possible for the person monitoring the call to record it offsite for later analysis. There are also systems that separately record or allow monitoring of the IVR and agent portions of the calls. Even without tracking the call to the end of interaction with an agent, SUI designers can glean much useful information from reviewing just the interaction with the IVR. The more calls the better (Suhm, 2008), but it is possible to get substantial and useful data from as few as 100 calls (Cohen et al., 2004). In addition to generating a list of usability problems discovered while reviewing recorded calls, the analysis of end-to-end calls can also provide the following information (Alwan & Suhm, 2010; Suhm, 2008):

- The call-reason distribution (for comparison with the distribution inferred from IVR reports and prioritization of changes)
- Call-path analysis (for identifying additional usability problems at the points in the call flow that have little traffic and high rates of unexpected abandonment or transfer, and for estimating the impact of design changes)
- Correct transfer rate (the rate at which callers transfer to the correct agent type)
- Total IVR benefit (an estimate of the agent time saved by the IVR per call)

Sample Design Documents

This section of the chapter contains sample design documents for a fictional application for a fictional company—Electric Razors International.

The sample documents include a high-level call flow, a detailed dialog specification, scenarios of use derived from the detailed dialog specification, and a recording manifest. The goal of the application is to provide self-service for locating stores that carry their products and for information about which razor heads go with which shaver. Callers who want to make a purchase transfer to an agent to handle that transaction, as long as the call center is open.

There are many ways to accomplish the goals of these sample design documents, but this is a set of SUI design documents that I have found useful in getting from concept to deployment. I should also point out that every time I return to this exercise, I find ways to improve the design, but, like any other project, this book has a deadline, and I must stop tweaking the design.

High-Level Call Flow

Figures 6.5 through 6.8 contain the high-level call flow. Figure 6.5 is an overview of the entire application. Figure 6.6 illustrates the call flow for a standard dialog step. Figure 6.7 shows more details about the process for finding a store, and Figure 6.8 has the call flows for replacement head information, transferring calls, and confirming call exit.

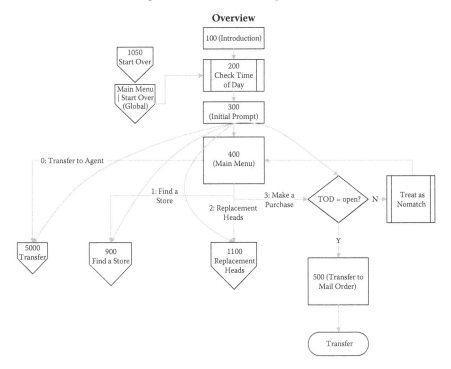

FIGURE 6.5
ERI IVR high-level call flow: Overview.

General Prompt Behavior

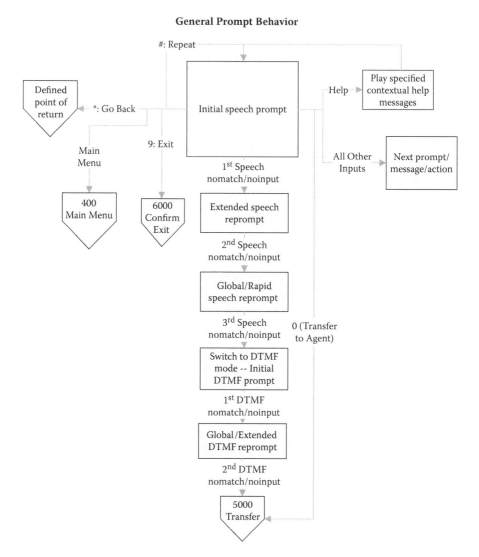

FIGURE 6.6
ERI IVR high-level call flow: General prompt behavior.

Detailed Dialog Specification (DDS)

Table 6.6 shows the DDS that supports the high-level call flow. The specification is organized by dialog step, with numbered steps that correspond to steps in the high-level call flow. The DDS contains almost everything that a programmer needs to code the call flow—the output audio and at least one expected user response, the names for audio and grammar files, connections among the dialog steps (From/To), system processing or conditions that affect the call flow, and design notes. When an element in the "System/Human" or

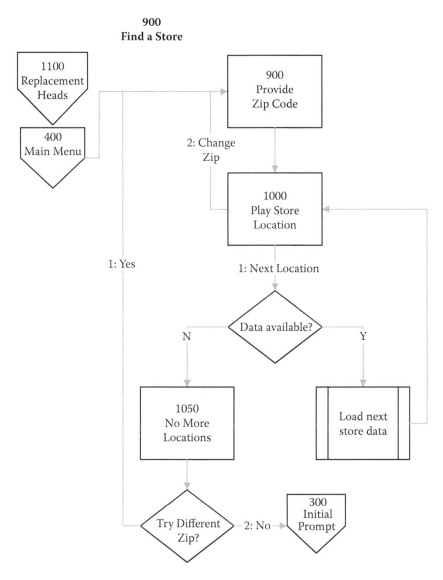

FIGURE 6.7
ERI IVR high-level call flow: Find a store.

"Message Type/DTMF Key" column changes, for example, from "System" to "Human" or "Speech Initial" to "Speech Help," the text is bold to make it easier to see.

The DDS typically underspecifies the design of the indicated grammars. Even a moderately complex grammar can interpret hundreds of different inputs, so the DDS provides one of the most expected utterances (and, if applicable, the associated DTMF touchtone key) for each option or input.

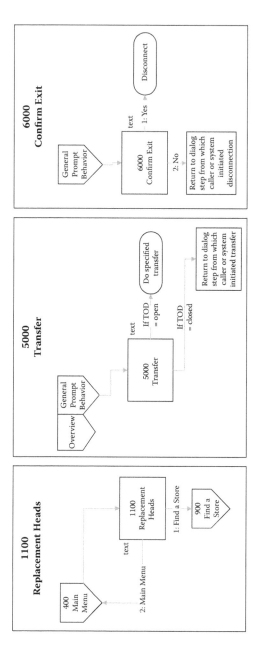

FIGURE 6.8
ERI IVR high-level call flow: Remaining dialog steps.

TABLE 6.6

Sample Detailed Design Document (DDS)

Row #	Step #	System/ Human	Message Type/DTMF Key	Prompts and Responses	Audio/Grammar Files	From/To	System Processing/ Conditions	Notes
2	100			**Introduction**		New Call		
3		**System**	**Message**	Welcome to the Electric Razor Customer Care Line.	intro			For definitions of Global Commands, see Step 2000 (Global Commands)
4		System	**Non-Speech Audio**	<AudioLogo based on branding tone>	audiobrandlogo			Audio tone based on company branding from other media
5		System	**Message**	Calls may be monitored and recorded.	callsmonandrec	200 (Check Time of Day)		Spoken softer and quicker
6								
7	**200**			**Check Time of Day**				
8		System	**Action**			100	If 7:59 am < Time < 8:01 pm, set TOD = open else TOD = closed	Check the time-of-day (TOD) to see if the call center is open or closed
9								

continued

TABLE 6.6 (continued)

Sample Detailed Design Document (DDS)

Row #	Step #	System/Human	Message Type/DTMF Key	Prompts and Responses	Audio/Grammar Files	From/To	System Processing/Conditions	Notes
10	300		**Initial Prompt**			200		
11		**System**	**Action**			400 (Main Menu)	If Mode = DTMF	If in DTMF mode go directly to Main Menu
12		System	**Speech Initial**	I can help you find a store, get replacement head info, or make a purchase. What would you like to do?	initial		If TOD = open AND Initial = true	This version plays when the call center is open and this is the first time through this dialog step
13		System	Speech Initial	I can help you find a store or get replacement head info. What would you like to do?	initial_ao		If TOD = closed and Initial = true	This version plays when the call center is closed and this is the first time through this dialog step
14		System	Speech Initial	Can I help you with anything else today?	noninitial		If Initial<>true	This plays on second and subsequent times through this dialog step
15		System	**Speech Help**			400 (Main Menu)		
16		System	**Speech Noinput/ Nomatch 1**			400 (Main Menu)		The silence timeout for this dialog step should be 3 s

#	State	Speaker	Mode	Text / Input	Grammar	Destination	Condition	Notes
17		**Human**						
18		Human		Find a store {store}	initial.grxml	900 (Find a Store)	Set Initial = false	
19		Human		Replacement heads {heads}	initial.grxml	1100 (Replacement Heads)	If TOD = open Set AgentPool = Orders; Set Initial = false	
20		Human		Make a purchase {purchase}	initial.grxml	500 (Transfer to Mail Order)	If TOD = closed Set Initial = false Treat as nomatch	
21		Human		Make a purchase {purchase}	initial.grxml		If TOD = closed Set Initial = false	
22		Human		Yes	initial.grxml	400 (Main Menu)	Set Initial = false	
23		Human		No	initial.grxml	2400 (Disconnect)	Set Initial = false	
24	0	Human		Transfer to Agent {transfer}	global.grxml	2200 (Transfer to Agent)	Set AgentPool = General	
25								
26	**400**	**System**		**Main Menu**		300		
27		System	**Speech Initial**	Please select:	pleaseselect			
28		System	Speech Initial	Find a Store,	findastore			
29		System	Speech Initial	Replacement Heads,	replacementheads			
30		System	Speech Initial	Make a Purchase,	makeapurchase		If TOD = open	This option only plays when the call center is open

continued

TABLE 6.6 (continued)

Sample Detailed Design Document (DDS)

Row #	Step #	System/ Human	Message Type/DTMF Key	Prompts and Responses	Audio/Grammar Files	From/To	System Processing/ Conditions	Notes
31		System	Speech Initial	or Exit.	orexit			
32		System	**Speech Help**	You're at the main menu.	atmain			The Speech Help audio plays if a caller says "Help"
33		System	Speech Help	To find the nearest store that carries our products, say Find a Store.	findastorehelp			
34		System	Speech Help	To find out what replacement head to buy for your razor, say Replacement Heads.	replaceheadhelp			
35		System	Speech Help	To order one of our products or accessories, say Make a Purchase.	purchasehelp		If TOD = open	
36		System	Speech Help	To end the call, say Exit.	exithelp			
37		System	Speech Help	<3 sec pause>	pause3000			
38		System	Speech Help	At any time you can say Repeat, Help, Go Back, Start Over, Transfer to Agent, or Exit.	always		If TOD = open	

#	Speaker	Type	Prompt	Name	Condition	Notes
39	System	Speech Help	At any time you can say Repeat, Help, Go Back, Start Over, or Exit.	always_ao	If TOD = closed	Don't offer Transfer to Agent if call center is closed—be responsive to it, but don't play it
40	System	Speech Help	<2 sec pause>	pause2000		
41	System	Speech Help	To continue,	tocontinue		
42	System	Speech Help	Please select:	pleaseselect		
43	System	Speech Help	Find a Store,	findastore		
44	System	Speech Help	Replacement Heads,	replacementheads		
45	System	Speech Help	Make a Purchase,	makeapurchase	If TOD = open	
46	System	Speech Help	or Exit.	orexit		
47	System	**Speech Noinput/ Nomatch 1**	To find the nearest store that carries our products, say Find a Store.	findastorehelp		
48	System	Speech Noinput/ Nomatch 1	To find out what replacement head to buy for your razor, say Replacement Heads.	replaceheadhelp		
49	System	Speech Noinput/ Nomatch 1	To order one of our products or accessories, say Make a Purchase.	purchasehelp	If TOD = open	
50	System	Speech Noinput/ Nomatch 1	To end the call, say Exit.	exithelp		

continued

TABLE 6.6 (continued)
Sample Detailed Design Document (DDS)

Row #	Step #	System/ Human	Message Type/DTMF Key	Prompts and Responses	Audio/Grammar Files	From/To	System Processing/ Conditions	Notes
51		System	**Speech Noinput/ Nomatch 2**	At any time you can say Repeat, Help, Go Back, Start Over, Transfer to Agent, or Exit.	always		If TOD = open	
52		System	Speech Noinput/ Nomatch 2	At any time you can say Repeat, Help, Go Back, Start Over, or Exit.	always_ao		If TOD = closed	
53		System	Speech Noinput/ Nomatch 2	<2 sec pause>	pause2000			
54		System	Speech Noinput/ Nomatch 2	To continue,	tocontinue			
55		System	Speech Noinput/ Nomatch 2	Please select:	pleaseselect			
56		System	Speech Noinput/ Nomatch 2	Find a Store,	findastore			
57		System	Speech Noinput/ Nomatch 2	Replacement Heads,	replacementheads			

#	Actor	State	Prompt	Name	Condition	Notes
58	System	Speech Noinput/Nomatch 2	Make a Purchase,	makeapurchase	If TOD = open	
59	System	Speech Noinput/Nomatch 2	or Exit.	orexit		
60	System	**Speech Noinput/Nomatch 3**	There seems to be a problem with this connection. For the rest of this call, please use the keypad.	switch	Set Mode = DTMF	Turn off speech reco and use DTMF for remainder of call
61	System	**DTMF Initial**	Main Menu.	mainland		
62	System	DTMF Initial	To find a store, press 1.	findstore_01		
63	System	DTMF Initial	For replacement heads, 2.	replacehead_02		
64	System	DTMF Initial	To make a purchase, 3.	purchase_03	If TOD = open	
65	System	DTMF Initial	To exit, 9.	exit_09		
66	System	**DTMF Noinput/Nomatch 1**	At any time you can press pound to repeat a menu, star to go back, or 0 to transfer to an agent.	always_dtmf	If TOD = open	
67	System	DTMF Noinput/Nomatch 1	At any time you can press pound to repeat a menu or star to go back.	always_dtmf_ao	If TOD = closed	
68	System	DTMF Noinput/Nomatch 1	<2 sec pause>	pause2000		

continued

TABLE 6.6 (continued)
Sample Detailed Design Document (DDS)

Row #	Step #	System/ Human	Message Type/DTMF Key	Prompts and Responses	Audio/Grammar Files	From/To	System Processing/ Conditions	Notes
69		System	DTMF Noinput/ Nomatch 1	To find the nearest store that carries our products, press 1.	findstorehelp_01			
70		System	DTMF Noinput/ Nomatch 1	To find out what replacement head to buy for your razor, press 2.	replacehelp_02			
71		System	DTMF Noinput/ Nomatch 1	To order one of our products or accessories, press 3.	purchasehelp_03			
72		System	DTMF Noinput/ Nomatch 1	To exit, press 9.	exithelp_09			
73		System	**DTMF Noinput/ Nomatch 2**	I'm sorry that I wasn't able to help you.	couldnthelp	2200 (Transfer to Agent)	Set AgentPool = General	
74								
75		**Human**	1	Find a store {store}	main.grxml	900 (Find a Store)		
76		Human	2	Replacement heads {heads}	main.grxml	1100 (Replacement Heads)		

77	Human	3	Make a purchase [purchase]	main.grxml	500 (Transfer to Mail Order)	If TOD = open Set AgentPool = Orders
78	Human	3	Make a purchase [purchase]	main.grxml		If TOD = closed Treat as nomatch
79	Human	9	Exit [exit]	global.grxml	2300 (Confirm Goodbye)	
80	Human	*	Go back [goback]	global.grxml	300 (Initial Prompt)	
81	Human	0	Transfer to Agent [transfer]	global.grxml	2200 (Transfer to Agent)	Set AgentPool = General
82						
83			**Transfer to Mail Order**		400	
500						
84	System	Message	Let me get someone to help with your order. Please hold.	mailorder		
85						
86						
900						
87	System	**Initial**	**Find a Store** OK, Store Locator.	storelocator	400, 1100	Play for both Speech and DTMF mode
88	System	**Speech Initial**	What's your ZIP code?	whatsyourzip		Enable DTMF entry here
89	System	**Speech Help**	You're at the store locator.	atstore		

continued

TABLE 6.6 (continued)
Sample Detailed Design Document (DDS)

Row #	Step #	System/ Human	Message Type/DTMF Key	Prompts and Responses	Audio/Grammar Files	From/To	System Processing/ Conditions	Notes
90		System	Speech Help	To find the nearest store that carries our products, please say or enter your five-digit ZIP code.	storehelp1			For your convenience you can find our products at the following retail stores: Walmart, Target, and Sears
91		System	Speech Help	<3 sec pause>	pause3000			
92		System	Speech Help	At any time you can say Repeat, Help, Go Back, Start Over, Transfer to Agent, or Exit.	always		If TOD = open	
93		System	Speech Help	At any time you can say Repeat, Help, Go Back, Start Over, or Exit.	always_ao		If TOD = closed	
94		System	Speech Help	<2 sec pause>	pause2000			
95		System	Speech Help	To continue,	tocontinue			
96		System	Speech Help	Please say or enter your five-digit ZIP code.	sayzip			

97	System	**Speech** **Noinput/** **Nomatch 1**	To find the nearest store that carries our products, please say or enter your five-digit ZIP code.	storehelp1
98	System	**Speech** **Noinput/** **Nomatch 2**	At any time you can say Repeat, Help, Go Back, Start Over, Transfer to Agent, or Exit.	always
99	System	Speech Noinput/ Nomatch 2	At any time you can say Repeat, Help, Go Back, Start Over, or Exit.	always_ao
100	System	Speech Noinput/ Nomatch 2	<2 sec pause>	pause2000
101	System	Speech Noinput/ Nomatch 2	To continue,	tocontinue
102	System	Speech Noinput/ Nomatch 2	Please say or enter your five-digit ZIP code.	sayzip
103	System	**Speech** **Noinput/** **Nomatch 3**	There seems to be a problem with this connection. For the rest of this call, please use the keypad.	switch Set Mode = DTMF

continued

TABLE 6.6 (continued)
Sample Detailed Design Document (DDS)

Row #	Step #	System/ Human	Message Type/DTMF Key	Prompts and Responses	Audio/Grammar Files	From/To	System Processing/ Conditions	Notes
104		System	DTMF Initial	For the nearest retail locations that carry our products, please enter your five-digit ZIP code.	enterzip			
105		System	DTMF Noinput/ Nomatch 1	At any time you can press pound to repeat a menu, star to go back, or 0 to transfer to an agent.	always_dtmf			
106		System	DTMF Noinput/ Nomatch 1	At any time you can press pound to repeat a menu or star to go back.	always_dtmf_ao			
107		System	DTMF Noinput/ Nomatch 1	<2 sec pause>	pause2000			
108		System	DTMF Noinput/ Nomatch 1	To find the nearest store that carries our products, please enter your five-digit ZIP code.	storehelp1_dtmf			
109		System	DTMF Noinput/ Nomatch 2	I'm sorry that I wasn't able to help you.	couldnthelp	2200 (Transfer to Agent)	Set AgentPool = Locator	

Line	Speaker	Input	Prompt	File	ID	Action
110						
111	**Human**		[Provides five digits]	zipcode.grxml	1000 (Play Store Location)	Set N = 1
112	Human		[Provides more or less than five digits]	zipcode.grxml		Treat like nomatch
113	Human	*	Go back [goback]	global.grxml	300 (Initial Prompt)	
114	Human	0	Transfer to Agent [transfer]	global.grxml	2200 (Transfer to Agent)	Set AgentPool = Locator
115						
116	**System**	**Initial**	**Play Store Location**		900, 1000	
117	System	Initial	The closest store is:	thestoreis		If N = 1
118	System	Initial	The next closest is:	thenextclosestis		If N > 1
119	System	Initial	<storeName>.	<computed>		<computed> means to play the desired audio file using information from backend database
120	System	Initial	Located at:	locatedat		
121	System	Initial	<address>.	<computed>		
122	System	Initial	Their telephone number is:	telephoneis		
123	System	Initial	<telephone number>.	<computed>		If N = 1
124	System	Initial	You can call them for business hours and local driving directions.	suggestyoucall		

(1000)

continued

TABLE 6.6 (continued)

Sample Detailed Design Document (DDS)

Row #	Step #	System/ Human	Message Type/DTMF Key	Prompts and Responses	Audio/Grammar Files	From/To	System Processing/ Conditions	Notes
125		System	**Speech Initial**	Please select Repeat, Next Location, Change Zip Code, Start Over, or Exit.	storemenu			
126		System	**Speech Help**	To repeat the information for this store, say Repeat. To hear information for another store near you, say Next Location. To enter a different ZIP code, say Change ZIP Code.	storemenuhelp1			
127		System	Speech Help	<3 sec pause>	pause3000			
128		System	Speech Help	At any time you can say Repeat, Help, Go Back, Start Over, Transfer to Agent, or Exit.	always		If TOD = open	
129		System	Speech Help	At any time you can say Repeat, Help, Go Back, Start Over, or Exit.	always_ao		If TOD = closed	
130		System	Speech Help	To continue,	tocontinue			

continued

131	System	Speech Help	Please select Repeat, Next Location, Change ZIP Code, Start Over, or Exit.	storemenu	
132	System	**Speech Noinput/ Nomatch 1**	To repeat the information for this store, say Repeat. To hear information for another store near you, say Next Location. To enter a different ZIP code, say Change ZIP Code.	storemenuhelp1	
133	System	**Speech Noinput/ Nomatch 2**	At any time you can say Repeat, Help, Go Back, Start Over, Transfer to Agent, or Exit.	always	If TOD = open
134	System	Speech Noinput/ Nomatch 2	At any time you can say Repeat, Help, Go Back, Start Over, or Exit.	always_ao	If TOD = closed
135	System	Speech Noinput/ Nomatch 2	To continue,	tocontinue	
136	System	Speech Noinput/ Nomatch 2	Please select Repeat, Next Location, Change ZIP Code, Start Over, or Exit.	storemenu	

TABLE 6.6 (continued)
Sample Detailed Design Document (DDS)

Row #	Step #	System/ Human	Message Type/DTMF Key	Prompts and Responses	Audio/Grammar Files	From/To	System Processing/ Conditions	Notes
148		Human	2	Change ZIP {changezip}	store.grxml	900 (Find a Store)		
149		Human	3	Main Menu {mainmenu}	global.grxml	200 (Check Time of Day)		
150		Human	9	Exit {exit}	global.grxml	2300 (Confirm Goodbye)		
151		Human	*	Go Back {goback}	global.grxml	900 (Find a Store)		
152		Human	0	Transfer to Agent {transfer}	global.grxml	2200 (Transfer to Agent)	Set AgentPool = Locator	
153								
154	1050	**System**	**Initial**	**No More Locations**		1000		
155		System	Initial	I'm sorry, the locations I've played for this ZIP code are the closest available.	closestplayed			
156		System	**Speech Initial**	Would you like to try a different ZIP code?	likediffzip			
157		System	Initial	<2 sec pause>	pause2000			
158		System	Speech Initial	Please say Yes, No, or Repeat.	yesnorep			
159		System	**Speech Help**	To try a different ZIP code, say Yes. To return to the main menu, say No.	difzip2			

continued

160	System	Speech Help	<3 sec pause>	pause3000	
161	System	Speech Help	At any time you can say Repeat, Help, Go Back, Start Over, Transfer to Agent, or Exit.	always	If TOD = open
162	System	Speech Help	At any time you can say Repeat, Help, Go Back, Start Over, or Exit.	always_ao	If TOD = closed
163	System	Speech Help	<2 sec pause>	pause2000	
164	System	Speech Help	To continue,	tocontinue	
165	System	Speech Help	Please say Yes, No, or Repeat.	yesnorep	
166	System	**Speech Noinput/Nomatch 1**	To try a different ZIP code, say Yes. To return to the main menu, say No.	difzip2	
167	System	**Speech Noinput/Nomatch 2**	At any time you can say Repeat, Help, Go Back, Start Over, Transfer to Agent, or Exit.	always	If TOD = open
168	System	Speech Noinput/Nomatch 2	At any time you can say Repeat, Help, Go Back, Start Over, or Exit.	always_ao	If TOD = closed
169	System	Speech Noinput/Nomatch 2	<2 sec pause>	pause2000	

TABLE 6.6 (continued)
Sample Detailed Design Document (DDS)

Row #	Step #	System/ Human	Message Type/DTMF Key	Prompts and Responses	Audio/Grammar Files	From/To	System Processing/ Conditions	Notes
170		System	Speech Noinput/ Nomatch 2	To continue,	tocontinue			
171		System	Speech Noinput/ Nomatch 2	Please say Yes, No, or Repeat.	yesnorep			
172		System	**Speech Noinput/ Nomatch 3**	There seems to be a problem with this connection. For the rest of this call, please use the keypad.	switch		Set Mode = DTMF	
173		System	**DTMF Initial**	To try a different ZIP code, press 1. To return to the main menu, press 2.	diizip2_dtmf			
174		System	DTMF Noinput/ Nomatch 1	At any time you can press pound to repeat a menu, star to go back, or 0 to transfer to an agent.	always_dtmf			
175		System	DTMF Noinput/ Nomatch 1	At any time you can press pound to repeat a menu or star to go back.	always_dtmf_ao			

176	System	DTMF Noinput/ Nomatch 1	<2 sec pause>	pause2000		
177	System	DTMF Noinput/ Nomatch 1	To try a different ZIP code, press 1. To return to the main menu, press 2.	difzip2_dtmf		
178	System	**DTMF Noinput/ Nomatch 2**	I'm sorry that I wasn't able to help you.	couldnthelp	2200 (Transfer to Agent)	Set AgentPool = Locator
179						
180	**Human**	1	Yes [yes]	yesno.grxml	900 (Find a Store)	
181	Human	2	No [no]	yesno.grxml	400 (Main Menu)	
182	Human	*	Go back {goback}	global.grxml	900 (Find a Store)	
183	Human	0	Transfer to Agent {transfer}	global.grxml	2200 (Transfer to Agent)	Set AgentPool = Locator
184						
185 **1100**	**System**	**Initial**	**Replacement Heads**			
186	System	Initial	If you have an IntelliTouch Razor, the replacement head is HZ8. For Ultima Razors, use UR9.	headsmessage	400	Play this whether in Speech or DTMF mode
187	System	**Speech Initial**	Please select Repeat, Find a Store, Go Back, Start Over, or Exit.	headsmenu		

continued

TABLE 6.6 (continued)
Sample Detailed Design Document (DDS)

Row #	Step #	System/ Human	Message Type/DTMF Key	Prompts and Responses	Audio/Grammar Files	From/To	System Processing/ Conditions	Notes
188		System	**Speech Help**	To repeat the information you just heard, say Repeat. To find a store where you can purchase this replacement head, say Find a Store.	headmodhelp1			
189		System	Speech Help	<3 sec pause>	pause3000			
190		System	Speech Help	At any time you can say Repeat, Help, Go Back, Start Over, Transfer to Agent, or Exit.	always		If TOD = open	
191		System	Speech Help	At any time you can say Repeat, Help, Go Back, Start Over, or Exit.	always_ao		If TOD = closed	
192		System	Speech Help	<2 sec pause>	pause2000			
193		System	Speech Help	To continue,	tocontinue			
194		System	Speech Help	Please select Repeat, Find a Store, Start Over, or Exit.	headmodmenu			

#	Speaker	Type	Prompt	Name	Condition
195	System	Speech Noinput/ Nomatch 1	To repeat the information you just heard, say Repeat. To find a store where you can purchase this replacement head, say Find a Store.	headmodhelp1	
196	System	Speech Noinput/ Nomatch 2	At any time you can say Repeat, Help, Go Back, Start Over, Transfer to Agent, or Exit.	always	If TOD = open
197	System	Speech Noinput/ Nomatch 2	At any time you can say Repeat, Help, Go Back, Start Over, or Exit.	always_ao	If TOD = closed
198	System	Speech Noinput/ Nomatch 2	<2 sec pause>	pause2000	
199	System	Speech Noinput/ Nomatch 2	To continue,	tocontinue	
200	System	Speech Noinput/ Nomatch 2	Please select Repeat, Find a Store, Start Over, or Exit.	headmodmenu	
201	System	Speech Noinput/ Nomatch 3	There seems to be a problem with this connection. For the rest of this call, please use the keypad.	switch	Set Mode = DTMF

continued

TABLE 6.6 (continued)

Sample Detailed Design Document (DDS)

Row #	Step #	System/ Human	Message Type/DTMF Key	Prompts and Responses	Audio/Grammar Files	From/To	System Processing/ Conditions	Notes
202		System	DTMF Initial	Press pound to repeat the information you just heard. To find a store where you can purchase this replacement head, press 1. To return to the main menu, 2. To end this call, 9.	headmod_dtmf			
203		System	DTMF Noinput/ Nomatch 1	At any time you can press pound to repeat a menu, star to go back, or 0 to transfer to an agent.	always_dtmf		If TOD = open	
204		System	DTMF Noinput/ Nomatch 1	At any time you can press pound to repeat a menu or star to go back.	always_dtmf_ao		If TOD = closed	
205		System	DTMF Noinput/ Nomatch 1	<2 sec pause>	pause2000			

continued

	Speaker	DTMF	Prompt / Response	Grammar	Destination	Action
206	System	DTMF Noinput/ Nomatch 1	Press pound to repeat the information you just heard. To find a store where you can purchase this replacement head, press 1. To return to the main menu, press 2. To end this call, press 9.	headmodhelp_ dtmf		
207	System	**DTMF Noinput/ Nomatch 2**	I'm sorry that I wasn't able to help you.	couldnthelp	2200 (Transfer to Agent)	Set AgentPool = Heads
208						
209	**Human**	#	Repeat {repeat}	global.grxml	1100 (Replacement Heads)	
210	Human	1	Find a store {findstore}	headmod.grxml	900 (Find a Store)	
211	Human	2	Main menu {mainmenu}	headmod.grxml	200 (Check Time of Day)	
212	Human	9	Exit {exit}	global.grxml	2300 (Confirm Goodbye)	
213	Human	*	Go back {goback}	global.grxml	400 (Main Menu)	
214	Human	0	Transfer to Agent {transfer}	global.grxml	2200 (Transfer to Agent)	Set AgentPool = Heads
215						

TABLE 6.6 (continued)

Sample Detailed Design Document (DDS)

Row #	Step #	System/ Human	Message Type/DTMF Key	Prompts and Responses	Audio/Grammar Files	From/To	System Processing/ Conditions	Notes
216	2000							**Global Commands**
217		Human		Help {help}		<specified in each step>	N/A	Play the designated speech help message. Note that the standard speech help message is the combination of the speech Nomatch/ Noinput messages, separated by 3 s of silence.
218		Human	#	Repeat {repeat}		<specified in each step>		Repeat the most recently played non-Help message (this is the default behavior for the built-in Repeat command in our VoiceXML browser), unless otherwise specified

No.	State	Speaker	Input	Prompt / Message	Grammar	Goto / Action	Comments
219		Human	*	Go Back {goback}		<specified in each step>	Return to the most immediately available previous dialog (specified for each dialog)
220		Human		Main menu {mainmenu}		200 (Check Time of Day)	This grammar includes "start over"
221		Human	0	Transfer to Agent {transfer}		2200 (Transfer to Agent)	
222		Human	9	Exit	Goodbye {exit}	2300 (Confirm Goodbye)	
223							
224	2200	System		**Transfer to Agent**	Global		
225		System	Message	Please hold for the next available agent.	transfer	If TOD = open	
226		System	Message	For quality purposes, your call may be monitored and recorded.	quality	If TOD = open Do the specified transfer	
227		System	Message	I'm sorry, our call center is currently closed. For help right now, visit our Web site, eri.com. If you need to speak with an agent, please call back during our business hours: 8 am to 8 pm, seven days a week.	canttransfernow	If TOD = losed Return to beginning of step in call flow from which transfer was initiated	

continued

TABLE 6.6 (continued)
Sample Detailed Design Document (DDS)

Row #	Step #	System/ Human	Message Type/DTMF Key	Prompts and Responses	Audio/Grammar Files	From/To	System Processing/ Conditions	Notes
228								
229	**2300**							
230		**System**	**Speech Initial**	**Confirm Goodbye** Do you want to end this call?	exit	Global		
231		System	**Speech Help**	To end the call, say Yes. To return to the system, say No.	exithelp1			
232		System	Speech Help	<3 sec pause>	pause3000			
233		System	Speech Help	At any time you can say Repeat, Help, Go Back, Start Over, Transfer to Agent, or Exit.	always		If TOD = open	
234		System	Speech Help	At any time you can say Repeat, Help, Go Back, Start Over, or Exit.	always_ao		If TOD = closed	
235		System	Speech Help	<2 sec pause>	pause2000			
236		System	Speech Help	To continue, please say Yes to end the call or No to return to the system.	exithelp2			
237		System	**Speech Noinput/ Nomatch 1**	To end the call, say Yes. To return to the system, say No.	exithelp1			

continued

238	System	**Speech Noinput/ Nomatch 2**	At any time you can say Repeat, Help, Go Back, Start Over, or Exit.	always	
239	System	Speech Noinput/ Nomatch 2	<2 sec pause>	pause2000	
240	System	Speech Noinput/ Nomatch 2	To continue, please say Yes to end the call or No to return to the system.	exithelp2	
241	System	**Speech Noinput/ Nomatch 3**	There seems to be a problem with this connection. For the rest of this call, please use the keypad.	switch	Set Mode = DTMF
242	System	**DTMF Initial**	To end the call, press 1. To return to the system, press 2.	exit_dtmf	
243	System	**DTMF Noinput/ Nomatch 1**	At any time you can press pound to repeat a menu or star to go back.	always_dtmf	
244	System	DTMF Noinput/ Nomatch 1	<2 sec pause>	pause2000	
245	System	DTMF Noinput/ Nomatch 1	To end the call, press 1. To return to the system, press 2.	exit_dtmf	

TABLE 6.6 (continued)

Sample Detailed Design Document (DDS)

Row #	Step #	System/ Human	Message Type/DTMF Key	Prompts and Responses	Audio/Grammar Files	From/To	System Processing/ Conditions	Notes
246		System	DTMF Noinput/ Nomatch 2	I'm sorry that I wasn't able to help you.	couldnthelp	2200 (Transfer to Agent)		
247								
248		**Human**	1	Yes	yesno.grxml	2400 (Disconnect)		
249		Human	2	No	yesno.grxml		Return to point in call flow from which user initiated disconnection.	
250		Human	*	Go Back	inline		Return to point in call flow from which user initiated disconnection.	
251		Human	0	Transfer to Agent [transfer]	global.grxml	2200 (Transfer to Agent)		
252								
253	**2400**	**System**	**Message**	**Disconnect**		2300		
254		**System**	**Message**	Thanks for calling the Electric Razor Customer Care Line.	thanksforcalling			
255		System	Message	You can find us on the Web at eri.com.	ontheweb			
256		System	Message	Goodbye!	goodbye		Disconnect	

Scenarios Derived from Detailed Dialog Specification

Tables 6.7 through 6.10 show four scenarios of use built using the dialogs specified in the DDS. One advantage of using a spreadsheet such as Excel to write the DDS is that you can link cells in the scenarios to cells in the DDS, so changes to those cells in the DDS automatically update the linked cells in the scenarios.

Recording Manifest

Table 6.11 shows the sample recording manifest, with the instructions I typically provide for the recording studio to use to trim the audio. Our experience has been that this works well to get audio in which the segments fit together well. The order of presentation corresponds to the order in the DDS because that provides a better understanding of context than, say, alphabetical order. More complex manifests include a column for direction notes, pronunciation guide, etc.

TABLE 6.7

Scenario 1: Caller Wants to Find the Nearest Store—Is Satisfied with the First Location

System:	Welcome to the Electric Razor Customer Care Line.
	<AudioLogo based on branding tone>
	Calls may be monitored and recorded.
	I can help you find a store, get replacement head info, or make a purchase. What would you like to do?
Caller:	Find a store.
System:	OK, Store Locator.
	What's your ZIP code?
Caller:	3 3 4 8 7.
System:	The closest store is:
	Fred's Razors.
	Located at:
	1111 Federal Highway, Boca Raton, Florida.
	Their telephone number is:
	561-555-1234.
	You can call them for business hours and local driving directions.
	Please select Repeat, Next Location, Change ZIP Code, Start Over, or Exit.
Caller:	Exit.
System:	Do you want to end this call?
Caller:	Yes.
System:	Thanks for calling the Electric Razor Customer Care Line.
	You can find us on the Web at eri.com.
	Goodbye!

TABLE 6.8

Scenario 2: Caller Wants to Find the Nearest Store—Is Satisfied with the Second Location

System:	Welcome to the Electric Razor Customer Care Line.
	<AudioLogo based on branding tone>
	Calls may be monitored and recorded.
	I can help you find a store, get replacement head info, or make a purchase. What would you like to do?
Caller:	Find a store.
System:	OK, Store Locator.
	What's your ZIP code?
Caller:	3 3 4 8 7.
System:	The closest store is:
	Fred's Razors.
	Located at:
	1111 Federal Highway, Boca Raton, Florida.
	Their telephone number is:
	561-555-1234.
	You can call them for business hours and local driving directions.
	Please select Repeat, Next Location, Change ZIP Code, Start Over, or Exit.
Caller:	Next location.
System:	The next closest is:
	Razor World.
	Located at:
	10017 Powerline Road, Boca Raton, Florida.
	Their telephone number is:
	561-555-9988.
	Please select Repeat, Next Location, Change ZIP Code, Start Over, or Exit.
Caller:	Exit.
System:	Do you want to end this call?
Caller:	Yes.
System:	Thanks for calling the Electric Razor Customer Care Line.
	You can find us on the Web at eri.com.
	Goodbye!

Availability of Sample Design Documents

You can never guarantee permanent availability of documents on the Web, but for the foreseeable future, you can download the following documents to use as design templates.

- High-level design documents (Figures 6.5 through 6.8)
  - Visio file: drjim.0catch.com/SampleHighLevelCallFlow_20100215.vsd
  - PDF file: drjim.0catch.com/SampleHighLevelCallFlow_20100215.pdf

TABLE 6.9

Scenario 3: Caller Wants to Find out Which Replacement Head to Buy for Razor

System:	Welcome to the Electric Razor Customer Care Line.
	<AudioLogo based on branding tone>
	Calls may be monitored and recorded.
	I can help you find a store, get replacement head info, or make a purchase. What would you like to do?
Caller:	Replacement heads.
System:	If you have an IntelliTouch Razor, the replacement head is HZ8. For Ultima Razors, use UR9.
	Please select Repeat, Find a Store, Go Back, Start Over, or Exit.
Caller:	Exit.
System:	Do you want to end this call?
Caller:	Yes.
System:	Thanks for calling the Electric Razor Customer Care Line.
	You can find us on the Web at eri.com.
	Goodbye!

TABLE 6.10

Scenario 4: Caller Wants to Purchase a Replacement Head (during Open Hours)

System:	Welcome to the Electric Razor Customer Care Line.
	<AudioLogo based on branding tone>
	Calls may be monitored and recorded.
	I can help you find a store, get replacement head info, or make a purchase. What would you like to do?
Caller:	Make a purchase.
System:	Let me get someone to help with your order. Please hold.

- Detailed design document (Tables 6.6-6.11)
  - Excel file: drjim.0catch.com/SampleDialog-20100215.xls

Summary

The typical SUI development cycle includes phases for concept, requirements, design, development, testing, deployment, and tuning. SUI designers typically have significant involvement in all phases except deployment.

Key SUI activities during the requirements phase are the analysis of users and tasks. Although there are many measurable individual differences among humans, the most important ones to consider for SUI design are age and expertise, both in the task domain and as a function of expected frequency of use of the speech application. If a large proportion of users will be

TABLE 6.11

Sample Recording Manifest

Count	Segment to Record	File Name
1	Welcome to the Electric Razor Customer Care Line.	intro
	<When editing the files, please use the following guidelines for trailing silence (the lead-ins should all be tightly trimmed with the exception of files starting with … which should have 500 ms of leading silence):	
	If the segment has no punctuation, trim the trailing edge tightly.	
	If the segment ends with a period or other sentence-final punctuation, leave 500 ms of silence at the trailing edge.	
	If the segment ends with a colon, leave 150 ms of silence at the trailing edge.	
	If the segment ends with any other type of punctuation, leave 250 ms at the trailing edge.	
	The tone for this application is friendly and energetic.>	
2	Calls may be monitored and recorded.	callsmonandrec
3	I can help you find a store, get replacement head info, or make a purchase. What would you like to do?	initial
4	I can help you find a store or get replacement head info. What would you like to do?	initial_ao
5	Can I help you with anything else today?	noninitial
6	Please select:	pleaseselect
7	Find a Store,	findastore
8	Replacement Heads,	replacementheads
9	Make a Purchase,	makeapurchase
10	or Exit.	orexit
11	You're at the main menu.	atmain
12	To find the nearest store that carries our products, say Find a Store.	findastorehelp
13	To find out what replacement head to buy for your razor, say Replacement Heads.	replaceheadhelp
14	To order one of our products or accessories, say Make a Purchase.	purchasehelp
15	To end the call, say Exit.	exithelp
16	At any time you can say Repeat, Help, Go Back, Start Over, Transfer to Agent, or Exit.	always
17	At any time you can say Repeat, Help, Go Back, Start Over, or Exit.	always_ao
18	To continue,	tocontinue
19	There seems to be a problem with this connection. For the rest of this call, please use the keypad.	switch
20	Main Menu.	mainland
21	To find a store, press 1.	findstore_01
22	For replacement heads, 2.	replacehead_02

TABLE 6.11 (continued)

Sample Recording Manifest

Count	Segment to Record	File Name
23	To make a purchase, 3.	purchase_03
24	To exit, 9.	exit_09
25	At any time you can press pound to repeat a menu, star to go back, or 0 to transfer to an agent.	always_dtmf
26	At any time you can press pound to repeat a menu or star to go back.	always_dtmf_ao
27	To find the nearest store that carries our products, press 1.	findstorehelp_01
28	To find out what replacement head to buy for your razor, press 2.	replacehelp_02
29	To order one of our products or accessories, press 3.	purchasehelp_03
30	To exit, press 9.	exithelp_09
31	I'm sorry that I wasn't able to help you.	couldnthelp
32	Let me get someone to help with your order. Please hold.	mailorder
33	OK, Store Locator.	storelocator
34	What's your ZIP code?	whatsyourzip
35	You're at the store locator.	atstore
36	To find the nearest store that carries our products, please say or enter your five-digit ZIP code.	storehelp1
37	Please say or enter your five digit ZIP code.	sayzip
38	For the nearest retail locations that carry our products, please enter your five-digit ZIP code.	enterzip
39	To find the nearest store that carries our products, please enter your five-digit ZIP code.	storehelp1_dtmf
40	The closest store is:	thestoreis
41	The next closest is:	thenextclosestis
42	Located at:	locatedat
43	Their telephone number is:	telephoneis
44	You can call them for business hours and local driving directions.	suggestyoucall
45	Please select Repeat, Next Location, Change ZIP Code, Start Over, or Exit.	storemenu
46	To repeat the information for this store, say Repeat. To hear information for another store near you, say Next Location. To enter a different ZIP code, say Change ZIP Code.	storemenuhelp1
47	For the Next Location, press 1. To Change the ZIP Code, 2. To return to the Main Menu, 3. To Exit, 9.	storemenu_dtmf
48	Press pound to repeat the information for this store. To hear information for another nearby store, press 1. To try a different ZIP code, press 2. To return to the main menu, press 3. To exit, press 9.	storehelp_dtmf
49	I'm sorry, the locations I've played for this ZIP code are the closest available.	closestplayed

continued

TABLE 6.11 (continued)

Sample Recording Manifest

Count	Segment to Record	File Name
50	Would you like to try a different ZIP code?	likediffzip
51	Please say Yes, No, or Repeat.	yesnorep
52	To try a different ZIP code, say Yes. To return to the main menu, say No.	difzip2
53	To try a different ZIP code, press 1. To return to the main menu, press 2.	difzip2_dtmf
54	If you have an IntelliTouch Razor, the replacement head is HZ8. For Ultima Razors, use UR9.	headsmessage
55	Please select Repeat, Find a Store, Go Back, Start Over, or Exit.	headsmenu
56	To repeat the information you just heard, say Repeat. To find a store where you can purchase this replacement head, say Find a Store.	headmodhelp1
57	Please select Repeat, Find a Store, Start Over, or Exit.	headmodmenu
58	Press pound to repeat the information you just heard. To find a store where you can purchase this replacement head, press 1. To return to the main menu, 2. To end this call, 9.	headmod_dtmf
59	Press pound to repeat the information you just heard. To find a store where you can purchase this replacement head, press 1. To return to the main menu, press 2. To end this call, press 9.	headmodhelp_dtmf
60	Please hold for the next available agent.	transfer
61	For quality purposes, your call may be monitored and recorded.	quality
62	I'm sorry, our call center is currently closed. For help right now, visit our Web site, eri.com. If you need to speak with an agent, please call back during our business hours: 8 am to 8 pm, seven days a week.	canttransfernow
63	Do you want to end this call?	exit
64	To end the call, say Yes. To return to the system, say No.	exithelp1
65	To continue, please say Yes to end the call or No to return to the system.	exithelp2
66	To end the call, press 1. To return to the system, press 2.	exit_dtmf
67	Thanks for calling the Electric Razor Customer Care Line.	thanksforcalling
68	You can find us on the Web at eri.com.	ontheweb
69	Goodbye!	goodbye

older users, then the system should accommodate their needs with regard to the pacing of speech and pauses. If many callers are or will become experienced, then the system should provide shortcuts tailored to their abilities. Designers should analyze tasks with the goal of achieving a high degree of efficiency, letting the system do extra work if it will reduce the number of steps required to complete a task. It is important to keep in mind that IVRs have two major self-service functions—task automation and accurate routing to agent skill groups—and that there can be significant benefits from partial automation where full automation is not possible.

During design, key activities are to develop the conceptual design, make high-level design decisions, make low-level design decisions, produce a high-level call flow, create a DDS, and prototyping. The primary goal of "Wizard of Oz" evaluation and other forms of prototyping is to support early usability testing.

Software developers do most of the development work, but SUI designers often have the responsibility of managing the audio recording process—selecting a studio, guiding the selection of a voice talent, producing the recording manifest, coaching the recording session, taking care of any final post-production audio edits, and delivering the audio to the software developers for deployment.

During testing, SUI designers may participate in the system verification test, the customer acceptance test, grammar testing, and formal usability testing. Ideally, usability testing of one sort or another takes place throughout the development cycle. Most often, SUI usability tests are primarily formative (seeking to discover and resolve usability problems) rather than summative (focused on achieving specific task-based measurements), although for projects with extensive iteration and access to data to use to establish quantitative goals, the later usability studies can be more summative.

In the tuning phase, activities that involve SUI designers are post-deployment surveys, call log analysis, and end-to-end call monitoring. With regard to call log analysis, it can be helpful if during the design phase there is a specification for each point in each dialog step regarding whether a hang-up or a transfer is expected (not indicative of a usability problem) or unexpected (indicative of a usability problem). Although call containment is a very common IVR metric, SUI designers should be aware of its serious limitations—failure to account for successful transfers and failure to account for the benefits of partial automation, which makes the IVR's performance appear worse than it really is, and failure to account for the hang-ups of frustrated callers, which makes the IVR's performance appear better than it really is—and should be ready to educate clients and others on the development team about its shortcomings. Finally, whenever the opportunity presents itself, SUI designers should perform end-to-end call monitoring. There is no substitute for hearing real callers trying to perform real tasks with the real speech application, and using the resulting data to drive improvements in SUI design.

References

Alwan, J., & Suhm, B. (2010). Beyond best practices: A data-driven approach to maximizing self-service. In W. Meisel (Ed.), *Speech in the user interface: Lessons from experience* (pp. 99–105). Victoria, Canada: TMA Associates.

ANSI. (2001). *Common industry format for usability test reports (ANSI-NCITS 354–2001).* Washington, DC: American National Standards Institute.

Balentine, B. (2007). *It's better to be a good machine than a bad person.* Annapolis, MD: ICMI Press.

Barr, R. A., & Giambra, L. M. (1990). Age-related decrement in auditory selective attention. *Psychology and Aging, 5*(4), 597–599.

Bloom, J., Gilbert, J. E., Houwing, T., Hura, S., Issar, S., Kaiser, L., et al. (2005). Ten criteria for measuring effective voice user interfaces. *Speech Technology, 10*(9), 31–35.

Cohen, M. H., Giangola, J. P., & Balogh, J. (2004). *Voice user interface design.* Boston, MA: Addison-Wesley.

Couper, M. P., Singer, E., & Tourangeau, R. (2004). Does voice matter? An interactive voice response (IVR) experiment. *Journal of Official Statistics, 20*(3), 551–570.

Denes, P. B., & Pinson, E. N. (1993). *The speech chain.* New York, NY: W. H. Freeman.

Dulude, L. (2002). Automated telephone answering systems and aging. *Behaviour and Information Technology, 21*(3), 171–184.

Dumas, J. S., & Salzman, M. C. (2006). Usability assessment methods. In R. C. Williges (Ed.), *Reviews of human factors and ergonomics* (Vol. 2) (pp. 109–140). Santa Monica, CA: Human Factors and Ergonomics Society.

Edlund, J., Gustafson, J., Heldner, M., & Hjalmarsson, A. (2008). Towards human-like spoken dialogue systems. *Speech Communication, 50,* 630–645.

Engelbrecht, K.-P., Quade, M., & Möller, S. (2009). Analysis of a new simulation approach to dialog system evaluation. *Speech Communication, 51,* 1234–1252.

Fisk, A. D., Rogers, W. A., Charness, N., Czaja, S. J., & Sharit, J. (2009). *Designing for older adults* (2nd ed.). Boca Raton, FL: CRC Press.

Fosler-Lussier, E., Amdal, I., & Kuo, H. J. (2005). A framework for predicting speech recognition errors. *Speech Communication, 46,* 153–170.

Gardner-Bonneau, D. (1999). Guidelines for speech-enabled IVR application design. In D. Gardner-Bonneau (Ed.), *Human factors and voice interactive systems* (pp. 147–162). Boston, MA: Kluwer Academic Publishers.

Gawron, V. J., Drury, C. G., Czaja, S. J., & Wilkins, D. M. (1989). A taxonomy of independent variables affecting human performance. *International Journal of Man-Machine Studies, 31,* 643–672.

Graham, G. M. (2005). *Voice branding in America.* Alpharetta, GA: Vivid Voices.

Graham, G. M. (2010). Speech recognition, the brand and the voice: How to choose a voice for your application. In W. Meisel (Ed.), *Speech in the user interface: Lessons from experience* (pp. 93–98). Victoria, Canada: TMA Associates.

Guinn, I. (2010). You can't think of everything: The importance of tuning speech applications. In W. Meisel (Ed.), *Speech in the user interface: Lessons from experience* (pp. 89–92). Victoria, Canada: TMA Associates.

Hajdinjak, M., & Mihelič, F. (2004). Conducting the Wizard of Oz experiment. *Informatica, 28,* 425–429.

Hajdinjak, M., & Mihelič, F. (2006). The PARADISE evaluation framework: Issues and findings. *Computational Linguistics, 32*(2), 263–272.

Halstead-Nussloch, R. (1989). The design of phone-based interfaces for consumers. In *Proceedings of CHI 1989* (pp. 347–352). Austin, TX: ACM.

Hone, K. S., & Graham, R. (2000). Towards a tool for the subjective assessment of speech system interfaces (SASSI). *Natural Language Engineering, 6*(3–4), 287–303.

Hura, S. L., Turney, S., Kaiser, L., Chapin, T., McTernan, F., & Nelson, C. (2008). The evolution of VUI design methodology. From the 2008 Workshop on the Maturation of VUI.

ISO. (2006). Common industry format (CIF) for usability test reports (ISO/IEC Standard 25062:2006). Available for order from http://www.iso.org/iso/iso_catalogue/catalogue_tc/catalogue_detail.htm?csnumber=43046

Kaiser, L., Krogh, P., Leathem, C., McTernan, F., Nelson, C., Parks, M. C., & Turney, S. (2008). Thinking outside the box: Designing for the overall user experience. From the 2008 Workshop on the Maturation of VUI.

Kelley, J. F. (1985). CAL—A natural language program developed with the OZ Paradigm: Implications for supercomputing systems. In *The First International Conference on Supercomputing Systems* (pp. 238–248). New York: ACM.

Larmon, C., & Basili, V. R. (2003). Iterative and incremental development: A brief history. *Computer, 36*(6), 47–56.

Larson, J. A. (2005). Ten guidelines for designing a successful voice user interface. *Speech Technology, 10*(1), 51–53.

Leppik, P. (2006). *Developing metrics part 1: Bad metrics. The Customer Service Survey.* Retrieved from www.vocalabs.com/resources/blog/C834959743/E20061205170807/index.html

Leppik, P., & Leppik, D. (2005). *Gourmet customer service: A scientific approach to improving the caller experience.* Eden Prairie, MN: VocaLabs.

Lewis, J. R. (1994). Sample sizes for usability studies: Additional considerations. *Human Factors, 36,* 368–378.

Lewis, J. R. (1996). Binomial confidence intervals for small sample usability studies. In G. Salvendy and A. Ozok (Eds.), *Advances in applied ergonomics: Proceedings of the 1st International Conference on Applied Ergonomics—ICAE '96* (pp. 732–737). Istanbul, Turkey: USA Publishing.

Lewis, J. R. (2000). *Developing minimal comprehensive sets of test cases for natural command grammars* (Tech. Rep. 29.3272, available at drjim.0catch.com/mintest-ral.pdf). Raleigh, NC: IBM Corp.

Lewis, J. R. (2001a). Evaluation of procedures for adjusting problem-discovery rates estimated from small samples. *International Journal of Human-Computer Interaction, 13,* 445–479.

Lewis, J. R. (2001b). Psychometric properties of the Mean Opinion Scale. In *Proceedings of HCI International 2001: Usability evaluation and interface design* (pp. 149–153). Mahwah, NJ: Lawrence Erlbaum.

Lewis, J. R. (2004). Effect of speaker and sampling rate on MOS-X ratings of concatenative TTS voices. In *Proceedings of the Human Factors and Ergonomics Society 48th annual meeting* (pp. 759–763). Santa Monica, CA: Human Factors and Ergonomics Society.

Lewis, J. R. (2006). Usability testing. In G. Salvendy (Ed.), *Handbook of human factors and ergonomics* (3rd ed.) (pp. 1275–1316). New York, NY: John Wiley.

Lewis, J. R. (2008). Usability evaluation of a speech recognition IVR. In T. Tullis & B. Albert (Eds.), *Measuring the user experience, chapter 10: Case studies* (pp. 244–252). Amsterdam, Netherlands: Morgan-Kaufman.

López-Cózar, R., Callejas, Z., & McTear, M. (2006). Testing the performance of spoken dialogue systems by means of an artificially simulated user. *Artificial Intelligence Review, 26,* 291–323.

López-Cózar, R., De la Torre, A., Segura, J. C., & Rubio, A. J. (2003). Assessment of dialogue systems by means of a new simulation technique. *Speech Communication, 40,* 387–407.

López-Cózar, R., De la Torre, A., Segura, J. C., Rubio, A. J., & Sánchez, V. (2002). Testing dialogue systems by means of automatic generation of conversations. *Interacting with Computers, 14,* 521–546.

Mayer, R. E. (1997). From novice to expert. In M. G. Helander, T. K. Landauer, and P. V. Prabhu (Eds.), *Handbook of human-computer interaction* (pp. 781–795). Amsterdam: Elsevier.

Möller, S., Engelbrecht, K., & Schleicher, R. (2008). Predicting the quality and usability of spoken dialogue services. *Speech Communication, 50,* 730–744.

Möller, S., & Skowronek, J. (2004). An analysis of quality prediction models for telephone-based spoken dialogue services. *Acta Acustica United with Acustica, 90*(6), 1112–1130.

Möller, S., Smeele, P., Boland, H., & Krebber, J. (2007). Evaluating spoken dialogue systems according to de-facto standards: A case study. *Computer Speech and Language, 21,* 26–53.

Nass, C., & Brave, S. (2005). *Wired for speech: How voice activates and advances the human-computer relationship.* Cambridge, MA: MIT Press.

Nass, C., Robles, E., Heenan, C., Bienstock, H., & Treinen, M. (2003). Speech-based disclosure systems: Effects of modality, gender of prompt, and gender of user. *International Journal of Speech Technology, 6,* 113–121.

Novick, D. G., Hansen, B., Sutton, S., & Marshall, C. R. (1999). Limiting factors of automated telephone dialogues. In D. Gardner-Bonneau (Ed.), *Human factors and voice interactive systems* (pp. 163–186). Boston, MA: Kluwer Academic Publishers.

Ogden, W. C., & Bernick, P. (1997). Using natural language interfaces. In M. Helander, T. K. Landauer, & P. Prabhu (Eds.), *Handbook of human-computer interaction* (pp. 137–161). Amsterdam, Netherlands: Elsevier.

Pashler, H., McDaniel, M., Rohrer, D., & Bjork, R. (2009). Learning styles: Concepts and evidence. *Psychological Science in the Public Interest, 9*(3), 105–119.

Polkosky, M. D. (2002). *Initial psychometric evaluation of the Pragmatic Rating Scale for Dialogues* (Tech. Rep. 29.3634). Boca Raton, FL: IBM.

Polkosky, M. D. (2006). Including the user in the conversation. In W. Meisel (Ed.), *VUI visions: Expert views on effective voice user interface design* (pp. 31–34). Victoria, Canada: TMA Associates.

Polkosky, M. D. (2008). Machines as mediators: The challenge of technology for interpersonal communication theory and research. In E. Konjin (Ed.), *Mediated interpersonal communication* (pp. 34–57). New York, NY: Routledge.

Polkosky, M. D. (2010). Opening the kimono. *Speech Technology, 15*(1), 7.

Polkosky, M. D., & Lewis, J. R. (2003). Expanding the MOS: Development and psychometric evaluation of the MOS-R and MOS-X. *International Journal of Speech Technology, 6,* 161–182.

Roscoe, S. N. (1997). The adolescence of engineering psychology. In S. M. Casey (Ed.), *The human factors history monograph series* (vol. 1) (pp. 1–9). Human Factors and Ergonomics Society: Santa Monica, CA.

Sadowski, W. J. (2001). Capabilities and limitations of Wizard of Oz evaluations of speech user interfaces. In *Proceedings of HCI International 2001: Usability evaluation and interface design* (pp. 139–142). Mahwah, NJ: Lawrence Erlbaum.

Sadowski, W. J., & Lewis, J. R. (2001). Usability evaluation of the IBM WebSphere "WebVoice" demo (Tech. Rep. 29.3387, available at drjim.0catch.com/vxml-live1-ral.pdf). West Palm Beach, FL: IBM Corp.

Sauro, J., & Lewis, J. R. (2005). Estimating completion rates from small samples using binomial confidence intervals: Comparisons and recommendations. In *Proceedings of the Human Factors and Ergonomics Society 49th annual meeting* (pp. 2100–2104). Santa Monica, CA: Human Factors and Ergonomics Society.

Sauro, J., & Lewis, J. R. (2009). Correlations among prototypical usability metrics: Evidence for the construct of usability. In *Proceedings of CHI 2009* (pp. 1609–1618). Boston, MA: ACM.

Sharit, J., Czaja, S. J., Nair, S., & Lee, C. (2003). Effects of age, speech rate, and environmental support in using telephone voice menu systems. *Human Factors, 45*, 234–251.

Sharma, C., & Kunins, J. (2002). *VoiceXML: Strategies and techniques for effective voice application development with VoiceXML 2.0.* New York, NY: John Wiley.

Sheeder, T., & Balogh, J. (2003). Say it like you mean it: Priming for structure in caller responses to a spoken dialog system. *International Journal of Speech Technology, 6,* 103–111.

Sinha, A. K., Klemmer, S. R., & Landay, J. A. (2002). Embarking on spoken-language NL interface design. *International Journal of Speech Technology, 5,* 159–169.

Stewart, R., & Wingfield, A. (2009). Hearing loss and cognitive effort in older adults' report accuracy for verbal materials. *Journal of the American Academy of Audiology, 20,* 147–154.

Suhm, B. (2008). IVR usability engineering using guidelines and analyses of end-to-end calls. In D. Gardner-Bonneau & H. E. Blanchard (Eds.), *Human factors and voice interactive systems* (2nd ed.) (pp. 1–41). New York, NY: Springer.

Suhm, B., & Peterson, P. (2002). A data-driven methodology for evaluating and optimizing call center IVRs. *International Journal of Speech Technology, 5,* 23–37.

Vredenburg, K., Mao, J. Y., Smith, P. W., & Carey, T. (2002). A survey of user-centered design practice. In *Proceedings of CHI 2002* (pp. 471–478). Minneapolis, MN: ACM.

Walker, M., Kamm, C., & Litman, D. (2000). Towards developing general models of usability with PARADISE. *Natural Language Engineering, 6*(3–4), 363–377.

Walker, M. A., Litman, D. J., Kamm, C. A., & Abella, A. (1997). PARADISE: A framework for evaluating spoken dialogue agents. In *Proceedings of ACL 97* (pp. 271–280). Madrid, Spain: MIT Press.

Walker, M. A., Litman, D. J., Kamm, C. A., & Abella, A. (1998). Evaluating spoken dialogue agents with PARADISE: Two case studies. *Computer Speech and Language, 12,* 317–347.

Watt, W. C. (1968). Habitability. *American Documentation, 19*(3), 338–351.

7

Getting Started: High-Level Design Decisions

Designing a SUI involves two levels of design decisions. First, you need to make certain high-level design decisions regarding system-level interface properties. Only then can you get down to the details of designing specific system prompts and dialogs. The high-level decisions include:

- Choosing the barge-in style
- Selecting recorded prompts or synthesized speech
- Deciding whether to use audio formatting
- Using simple or complex speech recognition
- Adopting a concise or verbose prompt style
- Allowing only speech input or speech plus touchtone
- Choosing a set of global navigation commands
- Deciding when and how to transfer to human agents
- Choosing a help mode or self-revealing help

There often is no single correct answer; the appropriate decisions depend on the application, the users, and the users' environment(s). The remainder of this chapter presents the trade-offs associated with each of these decisions.

Choosing the Barge-In Style

Enabling barge-in allows callers to interrupt system prompts as they play. In current speech applications, it has become very common to enable barge-in. Older applications could not permit barge-in, and thus required specialized user interface design. For information on the special design considerations for these older technologies (technically known as "half-duplex" systems), see Balentine, Ayer, Miller, & Scott (1997) and Balentine & Morgan (2001).

VoiceXML-compliant speech recognizers allow two different types of barge-in: hotword and speech. The different barge-in types differ in the type of audio input that will stop system speech. With speech barge-in, system speech stops as soon as the recognizer detects incoming speech or speech-like sounds. With hotword barge-in, the system continues speaking unless it detects that the caller has spoken a valid word or phrase in a currently active grammar. To choose the appropriate barge-in style, it is important to understand how the different styles interact with Lombard speech and the "stuttering effect."

Lombard Speech and the "Stuttering Effect"

When speaking in noisy environments, people tend to exaggerate their speech or raise their voices so others can hear them over the noise, a distorted speech pattern known as Lombard speech (Lombard, 1911). It can occur even when the only noise is the voice of another participant in the conversation (for example, when one person tries to interrupt another, or, in the case of a voice application, when the caller barges in while the computer is speaking).

The "stuttering effect" (Balentine & Morgan, 2001) can occur when a prompt continues playing for more than about 300–500 ms after the caller begins speaking. This is similar to the time (about 500 ms) required to resolve initiative conflicts in human–human dialogs (Schegloff, 2000; Yang & Heeman, 2010). Unless callers have undergone training with the system, they are likely to interpret the continued playing of the prompt as evidence that the system did not hear them. In response, some callers stop what they were saying and begin speaking again—causing a stuttering effect. It is often impossible for the system to match a stuttered utterance to anything in an active grammar, so the system identifies the input as out-of-grammar, even if what the caller intended to say was in an active grammar. This can then lead to cascading usability problems as the caller and system get out of sync.

Comparing Barge-In Detection Methods

Most current speech recognizers set for speech barge-in do a good job of detecting speech and stopping the production of audio in less than 500 ms. Thus a key advantage of speech barge-in is that its behavior is more consistent with typical human conversation, minimizing Lombard speech and the stuttering effect. Its disadvantage is its susceptibility to accidental interruption due to background noise, nonspeech vocalizations, and speech not intended for the system. The advantage of hotword barge-in is its resistance to accidental interruptions, such as those caused by coughing, muttering, or using the system in an environment with loud ambient conversation or other background noise. Its primary disadvantage is increased incidence of Lombard speech and the stuttering effect. Due to the potentially devastating consequences of the stuttering effect (and, to a lesser extent, of Lombard speech) on SUI usability, designers should use speech barge-in for their SUI applications.

If there is a requirement for the application to work in conditions of high ambient noise with untrained callers, it might be necessary to use hotword rather than speech barge-in. In that case, it is very important to design the application to minimize the need to barge in. Indeed, even when a system permits barge-in, many callers do not like to interrupt the system (Suhm, 2008). One way to ensure that the prompt stops before it can trigger a barge-in related stuttering effect is to keep the prompt short, decreasing the likelihood that the caller would have a reason to barge in. Another strategy is to promote short caller responses, ideally, no more than two or three syllables, because the average time required to produce a syllable of speech is about 200 ms (Crystal & House, 1990; Massaro, 1975). Finally, liberal use of brief pauses gives callers opportunities to begin talking without actively interrupting the system.

Selecting Recorded Prompts or Synthesized Speech

Synthesized speech (text-to-speech, or TTS) is useful as a placeholder during application development, or when the data to be spoken is "unbounded" (not known in advance), which makes it impossible to prerecord. When deploying speech applications, however, you should plan to use professionally recorded prompts whenever possible. Listeners generally prefer recorded natural speech to synthetic speech (Lines & Hone, 2002; Stern, Mullennix, & Yaroslavsky, 2006; Stevens, Lees, Vonwiller, & Burnham, 2005; Wang & Lewis, 2001), and there is experimental evidence that natural speech is easier for listeners to process mentally (Bailly, 2002; Francis & Nusbaum, 1999). Callers expect commercial systems to use high-quality recorded speech (Rolandi, 2004a), and even though current TTS has much better quality than the systems of 10 years ago, it is still true that only recorded speech can guarantee highly natural pronunciation and prosody.

Creating Recorded Prompts

For high-quality recorded prompts (Balentine & Morgan, 2001; Byrne, 2004; Cohen, Giangola, & Balogh, 2004; Larson, 2005):

- Use (typically by hiring through a studio) professional voice talent, quality recording equipment, and a suitable recording environment. Plan to call into and coach during recording sessions. Coaches must know the application, be familiar with the target voice attributes, have a good ear for subtle voice differences, and be able to verbalize the target voice attributes to the voice talent.
- Ensure consistency in microphone placement and recording area between recording sessions.

- Coach for natural rather than excessively "professional" speaking style.
- If prompts or messages contain long numbers, chunk the output in accordance with established conventions (for example, for phone numbers or Social Security numbers). If there are no conventions, group numbers into chunks of two or three digits (Dialogues Spotlight Research Team, 2000).
- If another speech segment will follow immediately, trim the beginning aggressively, but leave silence at the end that is appropriate for the ending punctuation (500 ms for final punctuation, 250 ms for non-final punctuation). Otherwise, leave as little silence as possible.
- As a general rule, use only one voice. When using multiple voices, have a clear design goal (for example, a female voice for introduction and prompts, and a male voice for menu choices).
- If a voice segment will appear in phrases with different intonations, be sure to record that segment for each intonation, and give each intonation its own file name. For example, suppose the system will seek confirmation of a telephone number using the phrase "Was that four three three <pause> five five six three?" The "three" that appears before the pause should have a slightly falling pitch, but the "three" that appears before the question mark should have a rising pitch. The "three" that appears between two other numbers should have a steady pitch. This suggests that to obtain the highest quality speech output, it will be necessary to obtain at least one recording for each of the three intonations. Note that the development effort required for this might not be appropriate for every application, with the final design decision balancing cost and quality requirements.
- Be aware of the appropriate stress to use in each planned audio segment. If the appropriate stress point is not the last open-class item (which is either a noun or a verb) in the sentence, make a note of where the speaker should place the stress.
- For segments that the application will play sequentially (in other words, will splice), be sure to choose the splice points carefully. If possible, choose splice points at natural pause points. Avoid splice points that separate articles such as "a," "an," and "the" from the following word (or any other combination that speakers normally run together or for which the correct article depends on the following word).
- If you intend to translate the application into other languages, plan ahead when defining the audio segments to record. If you do not speak the language, seek assistance from a native speaker. In general, try to avoid defining audio segments to record that are isolated nouns because in many languages the correct form for determiners (for example, in English, "a," "an," and "the"; in Spanish, "la" and "el") depends on the following noun. Also, there might be

other contextual dependencies that are important in the target language. Some of the known issues are gender sensitivity, ordering of recorded segments (for example, in collecting or playing street addresses), and plurality. Good planning early in the definition of audio segments can prevent unnecessary rework during translation (Ahlén, Kaiser, & Olvera, 2004).

Using TTS Prompts

Although recorded prompts are best for many applications, it is easier to maintain and modify an application that uses TTS prompts, making TTS ideal for use during development. Chapter 2 contains a description of TTS technologies. Here the focus is on using TTS in SUI design. Two areas of research in TTS usage for IVRs are (1) mixing recorded speech and TTS and (2) pronouncing proper names.

Spiegel (1997) conducted an experiment to see if the naturalness of recorded speech for prompts and the fixed portions of messages would overcome the well-known discomfort created by the discontinuities that occur when mixing natural and synthetic speech. The TTS in this experiment was formant, not concatenative, and only a male voice. The stimuli in the experiment included carrier phrases and variable information (shown in angle brackets), such as, "The number, 555-1234, is listed to <Fred's Happytime Bar>. The address is <123 Main Street>." The experiment had four conditions: M (recorded male voice for greetings and carrier phrases), M + S (recorded male speech for greetings, synthetic male speech for carrier phrases), F (female recordings), and F + S. Although the performance results were complicated, the rating results from the 62 participants indicated a strong caller preference for using the recorded male voice for greetings and carrier phrases—in other words, as much natural speech as possible with a consistent gender for the recorded and TTS voices. Callers particularly disliked the mixing of the recorded female voice with the male TTS.

McInnes, Attwater, Edgington, Schmidt, & Jack (1999) compared 10 ways to generate spoken output for a flight information service, ranging from the natural speech of a single speaker through various combinations of speakers and TTS to TTS only. In this experiment, the TTS was concatenative, based on the voice of one of the professional speakers. An example of the variable information (enclosed in angle brackets) and carrier phrases used in the experiment is: "Your flight from <Edinburgh> will leave on <Friday the second of August> at <oh eight thirty>, and will arrive at <London Heathrow> at <oh nine forty five>. From there it will depart for the onward journey at <eleven thirty>, arriving at <Vancouver> at <thirteen fifty>." This is a fairly rapid switching between variable information and carrier phrases. The results from 100 participants indicated a strong preference for using as much of the preferred professional voice as possible and for a professional over an amateur human speaker.

Gong, Nass, Simard, & Takhteyev (2001) framed the experimental issue as one between independent maximization (using as much natural speech as possible) and consistency (using all TTS to avoid juxtapositions between natural and synthetic voices). The 24 participants in the experiment heard seven sentences (about 2 min of speech) about university housing, for example, "If you are a graduate student, your chance of getting your first choice should be at least <five percent>." Half of the participants heard the sentences presented entirely in a male concatenative TTS voice; the other half heard the first part of the sentence spoken in recorded male natural speech (a graduate student, not a professional voice) and the variable portion (in angle brackets) in TTS. Participants rated the all-TTS condition more favorably than the mixed condition—better liked, more trusted, and more competent.

Gong and Lai (2003) used a "Wizard of Oz" methodology to study a voice-access-to-e-mail application. Participants completed eight tasks using this system and, depending on the condition, were exposed to all TTS or a mixture of TTS and recorded audio. In the mixed condition the dynamic content (e.g., the e-mail header and body) played in TTS, so there were probably few juxtapositions between recorded speech and TTS. Independent raters judged the all-TTS group to outperform the mixed group on the task scenarios; however, the all-TTS group had to listen to messages and calendar listings significantly more times to complete the tasks. Further, participants in the mixed condition judged their own performance to be better and judged the system to be easier to use than those in the all-TTS condition.

Lewis, Commarford, & Kotan (2006) had 72 randomly selected U.S. IBM employees listen to the following message, with variable information enclosed in angle brackets: "OK, for a <mid-sized nonsmoking car>, the rate is guaranteed <for one week> at <$39.99> per day minus ten percent through Triple-A, which lowers the rate to <$35.99> per day, with no drop charge. There is a 7 percent tax and a <$1.95 per day> vehicle licensing fee, for an approximate daily total of <$39>, with unlimited mileage." The TTS was a concatenative male voice that produced the audio for the entire message in the all-TTS condition and just the variable text in the mixed condition. The male speaker for the recorded segments was not a professional voice talent. All participants heard both versions of the message, half hearing the all-TTS version first.

Fifty-three participants (73.6%) indicated that they preferred the all-TTS presentation style, and 19 (26.4%) indicated a preference for the mixed presentation style ($t(71) = 14.07$; $p < 0.0005$). An adjusted-Wald 95% binomial confidence interval (Sauro & Lewis, 2005) for the percentage preferring all TTS ranged from 62.4% to 82.5%. Further an independent samples t-test indicated a significant main effect of order of presentation ($t(70) = 3.09$; $p < 0.0005$) such that participants were more likely to prefer the second presentation style that they heard. Those who listened to the mixed style first were more likely to prefer all TTS (89%) than those who listened to all TTS first (58%).

Consistent with the preference ratings, the all-TTS sample received more than twice as many positive comments as the mixed sample. Participants gave positive comments about the consistency of using just one voice (22.2% of the total number of all-TTS comments), and the smooth flow of the information (12.2%). Negative comments for the all-TTS condition included not liking its mechanical sound (25.6%) or lack of inflection (14.4%). The most cited negative trait of the mixed sample was the way the disjointed transitions between the voices distracted users from the information content (30.8%). Participants also complained about the large contrast between the sound of the recorded and TTS voices (18.0%) and the mechanical sound of the mixed sample (9.0%). Positive comments about the mixed presentation style included the pleasant sound of the recorded voice (9.0%) and the way the switch from one voice to the other distinguished the prices from the other information (7.7%). Interestingly, these participants felt that the switch from recorded to TTS called attention to the content, in contrast to the four times as many who reported that the switch distracted them from the content.

In general, participants disliked the degree of contrast between the voices in the mixed sample and liked the consistency of the all-TTS sample. They did not like the lack of inflection in the TTS voice or the mechanical sound of both the mixed and all-TTS samples.

The practice of mixing recorded audio and TTS between sentences in a single application is widely accepted and, in fact, necessary if one wishes to use recorded audio at all in an application that requires some TTS. In other words, if designers were to never mix, then all applications incorporating unpredictable information would necessarily play in all TTS. Therefore designers often use TTS "when they have to."

This collection of results is difficult to translate into recommendations for practice. The span covers 10 years, including years in which TTS technologies underwent considerable improvement. Adding to this difficulty is the lack of access to the specific audio used in the experiments—all TTS systems are not equal, and neither are all natural voices. Some experiments had professional human voices; others had amateur. Some stimuli included only a few juxtapositions between recordings and TTS; others had many. Table 7.1 summarizes the key experimental variables and results for these five experiments.

Although there are too few experiments to justify a firm conclusion, it is interesting that the only common experimental variable in the experiments in which participants preferred all TTS is that of professional voice. Specifically, the recorded voices in Gong et al. (2001) and Lewis et al. (2006) were amateur rather than professional.

Despite the difficulty of reconciling these disparate experimental findings, some guidelines consistent with the general findings are:

- For carrier phrases, use a professional rather than an amateur voice.
- Script messages to minimize the number of juxtapositions between recordings and TTS.
- Try to keep juxtapositions at natural pause points between phrases so the pauses can provide a buffer at the point of juxtaposition.

TABLE 7.1

Key Experimental Variables and Results for Mixing Recorded Speech and TTS

Study	TTS Technology	TTS Gender	Professional Voice	Voice Gender	Juxta-positions	Participant Preference
Spiegel (1997)	Formant	M	Yes	M/F	Few	Mixed (same gender)
McInnes et al. (1999)	Concatenative	F	Yes/No	F	Many	Mixed (professional)
Gong et al. (2001)	Concatenative	M	No	M	Few	All-TTS
Gong & Lai (2003)	Concatenative	M	Yes	M	Few	Mixed
Lewis et al. (2006)	Concatenative	M	No	M	Many	All-TTS

- Minimize juxtapositions by providing recorded speech for bounded variable information, with TTS fallback for unrecorded data.

If the information that the application needs to speak is unbounded, TTS is the only design choice. Examples of unbounded information include:

- Telephone directories
- E-mail messages
- Up-to-the-minute news stories
- Frequently updated lists of employee or customer names, movie titles, or other proper nouns

Enterprises often express a desire to use customers' names in their IVR dialogs, just as they do in scripting their interactions between customers and customer service representatives. It is, however, difficult to create and maintain recordings of names—in the United States there are more than 2,000,000 surnames and more than 100,000 first names (Spiegel, 2003a), making it attractive to produce proper names using TTS. Some of the difficulties English TTS engines have, however, when interpreting proper names (Damper & Soonklang, 2007; Henton, 2003; Spiegel, 2003a, 2003b) are:

- Uncommon patterns of English letter sequences
- Names with letter sequences that never appear in standard English, sometimes borrowed from other languages
- Names that have the same spelling but different preferred pronunciations
- Product names that contain symbols other than English letters

I know of no research on the magnitude of the adverse social effects of mispronouncing customer names, but mispronunciation would certainly do nothing to improve the relationship between a customer and an enterprise. Estimates of pronunciation accuracy for proper names produced by general (untuned) TTS engines range from about 50% (Henton, 2003) to 60–70% (Damper & Soonklang, 2007) to 70–80% (Spiegel, 2003b). After a 15-year research effort to improve proper name pronunciation, Spiegel (2003a, 2003b) described a tuned system that had 99% correct pronunciation for common names and 92–94% for uncommon names, demonstrating that the problem of accurate proper name pronunciation, although difficult, is possible to solve.

The implications of this for practical SUI design are:

- Unless an application requires it, avoid designs that address the caller by name.
- If you must address the caller by name and you do not want to mispronounce it, be prepared to make a substantial investment in crafting this part of the design.

SUI Personality and Persona

There is general agreement among SUI designers that applications that speak to users elicit automatic reactions that lead users to imply personality characteristics of the artificial speaker (Dahl, 2006; Larson, 2005). Most IVRs designed to provide service to callers should have a voice that is generally friendly and moderately enthusiastic without going "over the top," although there will be exceptions (for example, imagine an application designed for children to call and interact with while waiting in line at a theme park). In accordance with the appropriate social role of an IVR and the known characteristics of expert customer service representatives (Balentine & Morgan, 2001; Suhm, 2008), IVRs should assume the caller is busy, be efficient when communication is good, be helpful when progress is slower, be polite, rarely apologize, and never blame the caller.

But is there a need to go beyond this to create a more compelling and successful caller experience? This question has been at the heart of the debate over the use of personae in SUI design for more than 10 years (Balentine, 2010). As Balentine & Morgan stated in 2001 (p. 251):

> A recent trend in the speech recognition community emphasizes what is called the persona as an essential part of a telephony application. A persona is distinguished from a personality in its emphasis on the fictional nature of the character. Promoters of this trend specify the persona thoroughly and painstakingly before approaching other details

of the application—implying a belief that the persona is the foundation for every other design element. In many cases, considerable time and expense are devoted to the persona, including focus groups, psychological profiles, and voice talent audition and selection.

One of the arguments in favor of developing a persona is corporate branding (Kotelly, 2003). Another is to help designers produce a consistent caller experience by drawing on the characteristics of the persona to guide design elements such as tone, accent, specific wording choices, and confirmation styles (Cohen et al., 2004). To achieve this coordination among all the members of the SUI design team, Cohen et al. recommended writing a persona definition document containing at least a biographical sketch and a list of vocal attributes, and optionally a photograph to "bring the persona into focus" (p. 82).

For example, Roberts, Silver, and Rankin (2004) described a research effort to evaluate and select an appropriate voice persona for a large U.S. telecommunications company. They started by examining market research reports on the perception of the enterprise's customers regarding their current voice services and meeting with the enterprise's branding and marketing organizations, with the goal to balance customer needs and corporate image. As a result of this exercise, they determined that the vocal characteristics of the persona should be trustworthy, reliable, friendly, approachable, sincere, responsive, innovative, knowledgeable, caring, professional, attractive, influential, clear, helpful, and have a pleasant accent. Fifteen enterprise stakeholders applied these criteria to select the best five of nine professional voices from a vendor.

The next step was to get feedback about the voices from three focus groups (89 participants in all), gathering data with both open-ended questions and rating scales. There were significant differences between the focus group participants and enterprise stakeholders with regard to the preferred voice. From the open-ended questions, the desirable vocal qualities gleaned from the focus group participants for the persona (those that received mention by more than 10% of participants) were clear (65%), pleasant (29%), friendly (26%), natural (26%), neutral accent (18%), female (17%), and concise (17%). None of the participants specified that the persona should have a male voice. Roberts et al. (2004) concluded, "Together, the Voice Persona and the more natural customer interactions will help to provide a more positive customer experience than exists today" (p. 716).

Relatively well-known named voice personae are Amtrak's Julie and Bell Canada's Emily. As reported in *The New York Times* (Hafner, 2004):

> She's adventurous and well educated, friendly but not cloying, and always there to take your call. Meet Emily, the automated agent that answers customer-service calls for Bell Canada. So eager was Bell Canada to infuse the system with a persona that when Emily was introduced

to customers nearly two years ago, she came complete with small-town roots and a history degree from Carleton University in Ottawa. Her biography said she played volleyball and had backpacked around Asia after college.

Providing this level of detail in the biography of a persona raises the question of exactly how playing volleyball or being a history major at Carleton can possibly affect IVR scripting decisions (Balentine, 2007; Balentine & Morgan, 2001; Dahl, 2006; Klie, 2007). The effort expended by Roberts et al. (2004) might have optimized the voice used in the application, but there is no evidence that callers focused on trying to accomplish tasks with an IVR would perform more poorly or even notice the voice if the application had used the least preferred of the top five professional voices. Other arguments against designs focused on personae are:

- Investments in SUI should be on the user and efficient task completion rather than a focus on crafting the persona—"The enterprise exhibits shock and dismay when callers reject the solution because no one could afford a simple usability test, and yet $100,000 and more went to a detailed analysis of her clothes, her marital status, and whether there's a mole on her left arm" (Balentine, 2007, p. 102).

- When things go wrong in the IVR, having a branded name can give callers "something to latch onto … and that personality becomes a negative reflection on the entire company, not just the IVR" (Klie, 2007, p. 24).

- It might look fine on paper, but actually experiencing an "overly animated, artificially enthused IVR" (Rolandi, 2003, p. 29) can quickly become annoying to callers.

Walter Rolandi, one of the most outspoken writers against excessive investment in persona development, noted in his article "The Persona Craze Nears an End" that inappropriate, overly emotive personae can establish unrealistic user expectations, distract callers from their tasks, and annoy callers. Even appropriate personae cannot "make a dysfunctional application functional, add value to a questionable value proposition, or make an unusable application more usable" (Rolandi, 2007, p. 9).

Although it is common for enterprises to announce significant savings or improvements in customer satisfaction when deploying a persona-based IVR, there are many things that happen when deploying a new IVR, making it difficult to establish the extent to which the persona elements of the IVR contributed to the IVR's success (Dahl, 2006; Roberts et al., 2004). For example (Hafner, 2004), "Belinda Banks, a senior associate director of customer care at Bell Canada, pointed to a rise in customer satisfaction since Emily's arrival, but acknowledged it had more to do with relief at not having to enter touchtone commands." As I write this in 2010, Julie is still at Amtrak, but Emily is no longer with Bell Canada.

For practical SUI design, the following advice from Polkosky (2005b, p. 25) seems appropriate:

> Be concerned about persona, but not too much. While it is true that the speech and linguistic characteristics of an interface are important to user perceptions, persona should have the appropriate perspective: it is a secondary consideration after catering to users' needs. Undoubtedly, the friendliness and naturalness of the system voice are important characteristics that need to be controlled and the prompts should convey helpfulness and politeness; but don't let a blind focus on these design issues lead you to neglect the more important design decisions that enable a clear, simple, and efficient UI based on user goals.

Deciding Whether to Use Audio Formatting

Audio formatting refers to the use of nonspeech audio in SUI design. Audio formatting generally uses earcons rather then auditory icons. Auditory icons use naturally occurring sounds to represent specific actions or functions in a user interface (Gaver, 1986). By contrast, earcons are more abstract, musical sounds designed in "a rhythmicized sequence of pitches" (motives) and organized into families; these sounds represent actions and functions in an interface (Blattner, Sumikawa, & Greenberg, 1989, p. 23). The fixed characteristics of a motive are rhythm and pitch. Timbre, register, and pitch are variable characteristics of a motive. Earcons can consist of a single element or combinations that form compound earcons of several different tones in a distinct rhythm.

Audio formatting can be a useful technique because it increases the bandwidth of spoken communication by using speech and nonspeech cues to overlay structural and contextual information on spoken output. It can be particularly valuable as a concise way to convey branding information (for example, Sprint's sound of a pin dropping). Audio formatting is analogous to the well-understood notion of visual formatting, where designers use font changes, indenting, and other aspects of visual layout to cue the reader to the underlying structure of the information being presented (Dobroth, 2000).

Based on a literature review of the function of nonspeech audio in speech recognition applications (Polkosky & Lewis, 2001), some audio formatting recommendations are:

- Consider using audio cues to focus or shift user attention (for example, alerts, or to signal changes in topic)
- Consider using audio logos or familiar sounds to enable recall of previous knowledge or associations (for example, branding tones)

- Consider using tones to signal relevant aspects of the application's state (for example, acceptance and processing of input for processing delays that exceed 4 to 8 seconds—see Fröhlich, 2005)
- When possible, use complex auditory stimuli (such as music) rather than repetitive simple tones when callers are waiting, either for extended system response times or when on hold (Fröhlich, 2005; Polkosky, 2001)

A disadvantage of using audio formatting is that there are not yet standard sounds for specific purposes. Effective audio formatting might require the efforts of an audio designer (analogous to a graphic designer for graphical user interfaces) to establish a set of pleasing and easily discriminated sounds for these purposes. When designing audio formatting, keep the tones short: tones can be as short as 50–75 ms, and should typically be no longer than 500–1000 ms (Balentine & Morgan, 2001). Shorter tones are generally less obtrusive, so callers are more likely to perceive them as useful rather than distracting.

As a criticism of the use of tones in speech recognition systems, you might hear someone say, "I don't beep when I talk." The implication is that there is no place in speech recognition applications for nonspeech audio. Interestingly, although I've heard enterprise stakeholders make this comment, I've never heard a user complain about the presence of well-crafted nonspeech audio. Real callers who are trying to accomplish real tasks appear not to be terribly concerned with whether the application conforms in every detail to a human–human conversation.

Using Simple or Complex Speech Recognition

Most current speech applications use grammar-based recognition (as opposed to the use of natural language understanding, also known as statistical language model-based recognition), although the use of statistical methods is on the rise, especially for call routing. Grammars can be very simple or extremely complex, and the most appropriate grammar often depends on the selected prompting style. The grammars for directed dialog prompts (for example, "Please select checking, savings, or money market.") can be much simpler than the grammars for open-ended prompts (for example, "I can help you make, change, or cancel reservations. What would you like to do?")

Simple grammars are easier to code and maintain. When used with properly worded prompts that tend to limit the way most callers respond, simple grammars can be very effective. When taken to the extreme, however, an overly simple grammar might impose excessive restrictions on what callers can say. For example, you could design an application that requires callers to respond to every prompt with either "Yes" or "No," but this would result in a very cumbersome caller interface for all but the simplest of applications.

Natural command grammars are finite state grammars (as opposed to systems using statistical language models) that can approach natural language understanding (NLU) in its lexical and syntactic flexibility. Using these more complex speech recognition methods enhances the ability of systems to understand what callers say and can increase dialog efficiency (Brems, Rabin, & Waggett, 1995). The disadvantages of more complex speech recognition techniques are that they are currently more time-consuming to build and more difficult to test and maintain (Boyce, 2008).

Evaluating the Need for Complex Speech Recognition

New SUI designers often start out by assuming that unless a system has true, statistical NLU or NLU-like capability, it will not be usable. This assumption is not correct (Callejas & López-Cózar, 2008). Well-designed prompts can focus caller input so a fairly small grammar has an excellent chance of matching the caller input, and with careful prompt design, can often do so while maintaining a natural feel to the user interface. That said, complex grammars and statistical language modeling (the basis of NLU) can be advantageous for NLU call routing and flattening forms, and have become increasingly popular.

Flattening Menus

Research in the usability of spoken menus has shown that, when possible, it is desirable to flatten menu structures (Commarford, Lewis, Al-Awar Smither, & Gentzler, 2008). Specifically, relative to a deeper menu structure for the same number of choices, a flatter menu structure has fewer levels with more options per level. An emerging area in which NLU applications have seen considerable success is NL call routing, which is the ultimate menu flattening method. Callers respond to open-ended prompts such as "How may I help you?" and the application uses a number of statistical models to interpret the caller's response and to route the call appropriately (Gorin, Riccardi, & Wright, 1997; Kuo, Siohan, & Olive, 2003; Lee, Carpenter, Chou, Chu-Carroll, Reichl, Saad, & Zhou, 2000; Suhm, Bers, McCarthy, Freeman, Getty, Godfrey, & Peterson, 2002). This approach is especially effective if there are many potential destinations and no way to efficiently or clearly group them into categories (Boyce, 2008; Suhm, 2004).

If an analysis of the IVR requirements indicates that there will be many routing destinations from the initial prompt (either routing to human agents or to self-service components), then, from a high-level perspective, consider using NL call routing to give callers a way to avoid traversing a traditional menu structure to get the desired result. Note, however, that there will often still be a need to design and implement the traditional menu structure for callers to use if they experience difficulty with the NL call router. As long as a significant number of callers are successful with the NL call router, the application will be more efficient than those using a traditional approach. You need to know in advance if the application will use this approach because it will affect the detailed dialog strategy. In Chapter 8, see "Making Nondirective Prompts Usable" for specific guidance on managing caller responses to these types of open-ended (nondirective) prompts, and "Constructing Appropriate Menus and Prompts" for more details about research and practice in the design of spoken menus.

Flattening Forms

Another common IVR task is the collection of the bits of information needed to complete a transaction such as a car rental, analogous to entering the data on a form. The traditional approach is to collect the bits of information (tokens) one at a time—the "deep" approach. Advanced NLU applications that can parse the caller input and extract multiple tokens (see Chapter 2) can flatten the process of completing transaction forms, rather than requiring callers to provide these tokens one at a time. For example, if a caller to a travel reservations application says, "Tell me all the flights from Miami to Atlanta for tomorrow before noon," there are six tokens: all (rather than one or two), flights (rather than bus trips or trains), Miami (departure point), Atlanta (destination), tomorrow (date), before noon (time).

To take full advantage of the increased efficiency that form flattening provides, at some point the system should encourage callers to provide input with multiple tokens. Therefore, appropriate prompts for a natural command or NLU application are quite different from appropriate prompts for simple grammar systems. One way to do this is to provide a nondirective prompt with self-revealing help messages that contain examples of valid multiple-token commands (see Chapter 8, "Making Nondirective Prompts Usable").

There has been a considerable amount of research in the usability of NL call routing, but little research in the usability of NLU form flattening. In practice, there are current debates regarding whether the benefits of form flattening overcome the costs. As discussed in Chapter 6 (Table 6.1, Figure 6.3), enterprises can overestimate the extent to which their callers provide multiple tokens in response to open-ended prompts. Weegels (2000), in a usability study of an NLU travel reservation application, noted that although the system could accept multi-token utterances, "in practice, subjects rarely introduced information beyond the information asked for by the systems" (p. 78). A few major financial institutions have tried and then abandoned this method, citing the difficulty of getting their callers to use the approach successfully, the expense of getting the application deployed, and the expense of maintenance of the statistical models. Commercial systems using this method have existed, so these types of applications are definitely possible. The jury is still out as to whether they can be practical.

TABLE 7.2

Different Prompts for a PIN

Prompt	Syllable Count	Estimated Duration (s)
In order to access your records, I need you to provide your personal identification number.	27	5.4
Let's log in! Please tell me your four-digit personal identification number.	21	4.2
What is your four-digit PIN?	7	1.4
State your PIN.	3	0.6
What's your PIN?	3	0.6

Adopting a Concise or Verbose Prompt Style

Table 7.2 contains five versions of a prompt (the first four from Rolandi, 2004b) for a personal identification number (PIN), arranged in descending order of verbosity (wordiness). For each prompt, the table includes the number of syllables in the prompt and an estimate of the time required to play the prompt, assuming a duration of about 200 ms per syllable (Crystal & House, 1990; Massaro, 1975). Although counting syllables manually is more time-consuming than using an automatic word count, syllable counts provide more accurate estimates of prompt durations.

The first two versions are quite verbose, both exceeding 20 syllables, with estimated durations of 5.4 and 4.2 seconds. The last two are concise, each with 3 syllables and an estimated duration of less than 1 second. "State your PIN." is very directive; "What's your PIN?" is more conversational.

Accepted standards for terminology, formality, and interaction style in spoken communication vary based on demographic factors such as culture and socioeconomic status, as well as the subject matter or purpose of the conversation. Research (Polkosky, 2005a) and practice (Balentine & Morgan, 2001; Hambleton, 2000; Rolandi, 2004a, 2004b; Suhm, 2008) generally support adopting a concise initial prompting style over a more verbose style, because:

- More concise prompts take less time, with correspondingly less demand on a caller's working memory and faster task completion times
- More concise prompts are generally more consistent with conversational expectation, as long as they are also consistent with the expected service provider role of the IVR

Note that the emphasis on concise prompting applies primarily to the initial prompt spoken by the application. Help prompts that follow initial prompts tend to be more verbose.

Keep in mind that "concise" does not equal "machinelike" (Byrne, 2003). Observation of professional call center agents shows that they tend to use concise prompting to capture the information they need when filling out forms for customer orders and reservations. You can recover a human feeling through the selection of a good voice talent and providing short but natural variation in the prompts (Polkosky, 2005a). For example:

System: Super Club auto reservations. You can use this system to make, change, or cancel reservations. What would you like to do?

Caller: I'd like to make a reservation.

System: For what city?

Caller: Pittsburgh.

System: Picking up on which day?

Caller: Thursday, May the third.

System: And your arriving airline and flight?

Caller: Delta 2329

A particularly advantageous way to provide variation is through the use of discourse markers such as "first," "next," "finally," and "now." Other useful discourse markers include "oh," "otherwise," "OK," and "sorry." Each of these words has its own function in linguistic discourse (see Chapter 3). To achieve a natural-sounding dialog, however, avoid overusing them.

Allowing Only Speech Input or Speech plus Touchtone

Mixed-mode applications are those that mix speech and touchtone (also referred to as Dual-Tone Multi-Frequency, or DTMF). Because speech applications and touchtone applications do not typically have the same "sound and feel," it can be tricky to mix the two, making this a high-level decision that will affect the content of prompts (see "Mixed Mode Prompting" in Chapter 8).

The primary advantage of a touchtone mode is that it allows callers to continue self-routing and self-service when speech doesn't work well (Balentine, 2010; Rolandi, 2004a; Suhm, 2008), for example, due to:

- Caller accent
- Speech disability
- High ambient noise
- Desire to enter sensitive information without speaking it (for example, a PIN)

It is possible to manage usability issues due to caller accent or speech disability by offering touchtone alternatives without disabling speech. Despite continuing research on improving speech recognition in adverse conditions (Karray & Martin, 2003), high ambient noise will cause problems unless the application disables speech barge-in. With speech barge-in enabled, high ambient noise (for example, the noise in an airport terminal) can cause the application to repeatedly stop playing prompts and messages, often almost as soon as they begin playing. Because disabling speech barge-in can cause significant usability problems with speech input, it is better to disable speech input altogether and to stay in touchtone mode following multiple speech failures, still permitting touchtone barge-in (Attwater, 2008). As McKellin, Shahin, Hodgson, Jamieson, and Pichora-Fuller (2007, p. 2170) noted with regard to human–human dialog, "Conversations do not necessarily succeed in noisy settings."

This can keep callers in the IVR who would otherwise have to speak to a representative whether or not they wanted to do so. When working out high-level design decisions, enterprises must balance the relative costs and benefits of offering a touchtone fallback versus faster transfer to agents. For example, enterprises that face significant competition and with customers who can easily change providers might choose a strategy of rapid transfer over touchtone fallback.

Choosing a Set of Global Navigation Commands

Should SUI designs include universal global navigation commands? Rosenfeld, Olsen, and Rudnicky (2001, p. 41) argued in favor of universal commands as providing a consistent "sound and say" analogous to consistent "look and feel" in graphical user interfaces. Currently there is little support in the industry for rigid standards for global navigation commands (Boretz, 2009), but there do appear to be a small set of global navigation functions that many speech-enabled IVRs support. Six of the most common global navigation commands (Cohen et al., 2004; Lewis, Simone, & Bogacz, 2000) appear in Table 7.3, along with their functional description and alternate commands (synonyms) suggested by research such as the Speech Graffiti Project (Shriver & Rosenfeld, 2002; Tomko, Harris, Toth, Sanders, Rudnicky, & Rosenfeld, 2005).

The first two, Repeat and Help, stay in the current dialog step, aiding the caller in completing the step. The second two, Go Back and Start Over, give callers the ability to get out of the current dialog step and move back, either to the immediately preceding step or to the beginning. The third two, Transfer to Agent and Exit, get callers out of the application, either transferring to a customer service representative or disconnecting the call.

TABLE 7.3

Common Global Navigation Commands

Command	Functional Description	Alternates
Repeat (#)	Repeats the most recently played information—always the most recently played prompt and any informative messages that played just before the prompt	"Repeat that," "What?"
Help	Plays a help message	"Options"
Go Back (*)	Backs up to the previous dialog step	"Back," "Undo," "Scratch that"
Start Over	Returns to the initial dialog step (if menu based, the main menu)	"Main menu," "Restart"
Transfer to Agent (0)	Transfers to a human call center agent or plays a message about the availability of agents	"Operator"
Exit	Ends the call, usually after confirmation of the intent to disconnect	"Goodbye"

There is no standard for mapping touchtone (DTMF) keys to these global functions. Table 7.3 includes a mapping for U.S. applications based on principles of control-display compatibility (* for Go Back, # for Repeat) and convention (0 for Transfer to Agent). There is no strong need to assign a key to the remaining global commands. Noinput and nomatch events trigger help messages, callers can hang up to disconnect, and once discovered, callers can repeat Go Back to get to any previous dialog step.

Go Back

Some speech applications provide Go Back as a global navigation command, but others do not. Including Go Back increases the cost of a speech application due to the need to decide, for each dialog step, exactly where Go Back will take the call flow, to code the Go Back function, and then to test it. This expense, plus evidence from some application log reports that relatively few callers use Go Back (Dougherty, 2010), has led some SUI designers to recommend against including Go Back in the global command set.

On the other hand, usability tests that I've conducted have led me to believe that as long as the global command set is discoverable, callers who need it will discover, appreciate and use the Go Back function. In any reasonably complex application, callers (especially first-time callers) might need to explore the interface. While exploring, they might go down an unintended path, and will need a command to back up through the menu (dialog) structure. For example:

System: Do you want to leave a message or forward the call?

Caller: Forward the call.

System: Forward to which four-digit extension?

TABLE 7.4

Usage of "Go Back" in a Usability Study of an
Information Providing IVR

Participant	Task 1	Task 2	Task 3	Task 4	Task 5
1		X		X	X
2	X				
3					
4				X	
5		X	X	X	
6					
7	X		X	X	

Caller: <Realizes that she doesn't know the extension, not sure what to do now>

System: <noinput timeout> At any time you can say Repeat, Help, Go Back, Start Over, Transfer to Agent, or Exit.

Caller: Go back.

System: Do you want to leave a message or forward the call?

I know of no published research on the topic of when to include Go Back as a global command in a speech recognition IVR. In a usability study I conducted on an application designed to provide different types of information (weather, sports, etc.), five of the seven participants discovered and used the Go Back command, some repeatedly, as shown in Table 7.4. On the other hand, none of the three participants in a very small-scale study of a bill-paying IVR used Go Back when completing their tasks—possibly because the success rate at each dialog step was very high and there was little need to navigate. It appears that the need to invest in Go Back might differ from application to application. Unless a SUI designer is very confident that callers will not need the ability to back up a step (either in a menu structure or in a form-filling procedure), the safer approach is to provide the global Go Back command.

Exit

Some callers just hang up when they have finished using an application. Other callers, however, are more comfortable taking an explicit application action (such as saying Exit or Goodbye) to end the call. For deployed systems, the reported percentage of callers taking explicit actions to end calls ranges from 10% to 15% (Callejas & López-Cózar, 2008; Turunen, Hakulinen, & Kainulainen, 2006). An advantage of designing for an explicit action is that this gives the application one more opportunity to communicate with callers, such as:

- Prompting the caller with an up-sell/cross-sell option
- Inviting the caller to participate in a post-call satisfaction survey

- At the very least, thanking the caller and, incidentally, providing the caller with a clear indication of task completion so there is no lingering uncertainty (and possibly identifying additional information required to complete the transaction)

For example:

Caller: Exit. <After having indicated the desire to make an appointment>

System: You haven't specified an appointment date. Do you want to cancel this appointment?

Caller: No.

System: OK. On which date would you like to make the appointment?

As with other global commands, if callers frequently trigger the Exit command accidentally due to acoustic confusability with other commands (such as "Next"), it might be necessary to reword the commands, focusing on changing the less-frequently used command and continuing to consider acoustic confusability. For example, it's common to include the word "Goodbye" as a synonym for "Exit"—but probably not in an application that includes the command "Reply" or the destination "Kuai" (Lai, Karat, & Yankelovich, 2008).

Also, you don't want to lose a call due to this type of misrecognition. One approach is to include confirmation for the exit action. For example:

Caller: Exit.

System: Are you sure you want to end this call?

Caller: Yes.

System: Thanks for calling. Goodbye.

An alternative approach is for the system to respond with "Goodbye" rather than an explicit confirmation, but to remain "active in case the input was misrecognized. If the user wants to continue, they can simply speak again; if not, they can just hang up" (Tomko et al., 2005, pp. 10–11). I know of no research on the relative effectiveness of these two approaches.

Using the Global Commands

Simply having these commands in the system is not sufficient to make them usable; callers need to know that they exist. Here are three ways to provide information about global commands:

1. At some point in the sequence of help prompts, tell the caller about the global commands. The global navigation commands are pretty straightforward, easily understood by callers without additional

explanation. For example, the following prompt seems to work well for reminding callers about the global navigation commands:

System: At any time you can say Repeat, Help, Go Back, Start Over, Transfer to Agent, or Exit.

2. List key global commands at task terminal points, either as the last choices in a short menu or following a 1500 to 2500 ms pause after the last option in a longer menu. A caller is most likely to want to navigate away at the end of a task so provide information about global commands the caller might need when the caller most needs them. For example:

System: You have transferred one thousand dollars from savings to checking. The confirmation number is 6 5 4 3 2 1. Select Repeat, Perform Another Transaction, Go Back, Main Menu, or Exit.

or:

System: You have received one new message from David Jones. Select Play Message, Reply, Reply to All, or Delete Message. <2000 ms pause> At any time you can say Repeat, Help, Go Back, Main Menu, Transfer to Agent, or Exit.

3. The application's introductory message can tell callers about the two or three most important global commands. Because including this message will increase the time it takes for callers to get to the first prompt, designers should not automatically use this strategy for informing callers about global commands, reserving its use for situations in which repeat callers will benefit enough from hearing this information early in a call to overcome its disadvantages (see Chapter 8, "Short List of Global Commands").

Deciding Whether to Use Human Agents in the Deployed System

A key motivation for developing interactive speech systems is to reduce the cost of call centers by handling routine calls automatically (self-service) and routing other calls to the correct agent skill group. Depending on an application's purpose, human agents either will or will not be available to callers. This decision has consequences for the design of a speech application.

For example, if agents are available, the set of global commands should include the Transfer to Agent command (and reasonable synonyms).

If high-quality customer service is a main goal of the application, then it must include the ability to transfer calls to human agents, either automatically on the detection of caller difficulty with the application or at the caller's request.

If reduced cost is the overriding goal of the application, there might not be any human agents. In this case, the Transfer to Agent command should be in a global grammar, but the system response on hearing the command would be to inform the caller that no agents are available. This would prevent the speech recognition engine from trying to match the caller input ("Transfer to Agent") with something else in the grammar, which could lead to misrecognition. It also allows the application to directly and unambiguously inform the caller that the system does not have any human agents and provides an opportunity to tell callers about other e-Service options such as a Web site.

Designing for In-Hours and After-Hours Operation

Unless the call center is always open, you'll need to decide whether to design your application to operate differently depending on whether the call center is open (in hours) or closed (after hours). This decision affects the content of global messages (for example, no mention of transfer to agent after hours, even though it should always be in the active grammar) and the modification of menus to eliminate direct-transfer options after hours.

The primary advantage of designing after-hours menus without direct-transfer options is to avoid annoying callers who have cooperated with the IVR along a routing path, only to find out after answering a series of questions that the call center isn't open. Also, focusing the IVR on self-service after hours generally leads to simpler menu structures to guide callers to the available self-service functions.

Managing Early Requests for and Transfers to an Agent

In general, applications should prompt for transfer to an agent only when there is clear evidence that the caller is experiencing difficulty (multiple requests for help, noinputs, and/or nomatch events). Automatically transfer to an agent only when the caller experiences repeated difficulty in responding to a dialog step. The leading practice is to NOT announce representative access up front, even if it is available. Balentine (2006) reported that in one case, briefly delaying the announcement of the availability of agents until after the main menu led to a 2% rise in call containment. "Callers who speak during the silence are given their self-service choice and succeed in high numbers. Callers who wait to hear the CSR prompt still have a better memory of the menu because of the pause" (Balentine, 2006, p. 90).

Even though the leading practice is to not announce agent access up front, callers might attempt to reach a representative as their first interaction with the IVR. If they do, it's important to recognize this and respond appropriately (Balentine, 2007; Larson, 2005). Failure to do so wastes time for both the caller and the enterprise. Rather than having the effect of increasing caller usage of the IVR, making it difficult to reach agents leads to caller behaviors such as hanging up and redialing, pressing 0 repeatedly, playing possum, coughing, or saying nonsense words—in other words, anything to break through the IVR and talk to an agent. When one of these strategies does work, the first thing many callers do is to complain to the agent about the difficulty of getting through, which wastes even more time (and even more expensive time, from the enterprise's point of view). If the caller finds it excessively difficult or impossible to reach a representative, then another strategy is to call competitors to see if they offer better customer service.

For these reasons, the leading practice is to respond immediately to requests for agents—indicating recognition of the caller's request. Some design options are:

- Transfer the caller as requested
- Tell the caller that no agents are available (e.g., after hours, holiday) and, ideally, offer a call back option
- If early in the call, negotiate with the caller to stay in the IVR

It seems reasonable to provide callers with value proposition such as:

System: For fastest service, please choose one of the following:

System: OK. To get you to the right person as quickly as possible, please choose one of the following:

Note that the second version implies a transfer rather than self-service, so use it only in hours, not after hours, or set a variable that will ensure an appropriate transfer at the earliest possible point in the interaction. When callers reject an attempt at negotiation, then it is probably best to transfer them immediately to a general agent skill group.

Hura (2008, p. 7) reported better success at getting callers to make an initial selection with "OK, I'll get you to an agent, but first please tell me if you need help with A, B, C, or D" than with "OK, I can transfer you to an agent after you make a selection." She interpreted the relative success of the first prompt as, "User in this case successfully and happily made appropriate selections and were almost always routed correctly; this prompt obviously motivates users to make a good choice because there is a direct benefit to them. The same benefit exists for the first prompt, but the wording makes the selection seem like just another hoop they must jump through for the sake of the automated system."

Choosing a Help Mode or Self-Revealing Contextual Help

Most applications include the command Help (and possibly a set of alternative phrases that act as synonyms for help) in their global grammars. But what should an application do when a caller asks for help? It can either switch to a separate help mode or use a technique known as self-revealing contextual help. Because help modes introduce the possibility of mode-related usability problems (need for mode-switching, confusion about current mode, etc.) and global help messages (messages that try to explain generally how to use the application rather than how to deal with the current dialog step) are rarely useful, it's better practice to use contextual help.

Maximizing the Benefits of Self-Revealing Contextual Help

It's possible that a caller might experience momentary confusion or distraction, which leads the caller to explicitly request help. If the system has been designed to be self-revealing, these explicit requests for help could receive the same treatment as a silence timeout or an out-of-grammar utterance, allowing you to reuse the same code and prompts. This is an implicit style of help because the system never enters an explicit help mode. Implicit help provides more of a sense of moving forward (resulting in a better, more satisfying user interface) and is simpler to code than an explicit help mode.

Implementing Self-Revealing Contextual Help

Note the self-revealing properties of the following dialog:

> System: Welcome to our automated directory dialer. You can call any of our employees by speaking the employee's name and location. You can say Help or Exit at any time.
>
> Caller: (interrupting): Help.
>
> System: To start, please say the desired name and location.
>
> Caller: [Silence timeout]
>
> System: For example, to call Joe Smith in Kansas City, say, "Joe Smith, Kansas City."
>
> Caller: [Caller coughs; system interprets as nomatch]
>
> System: At any time you can say Repeat, Help, Go Back, Start Over or Exit. To continue, please say the desired name and location.
>
> Caller: Ed Black, Poughkeepsie.

Notice the self-revealing nature of the dialog, and the way it can work to keep the caller moving forward whether triggered by a silence timeout, an utterance from the Help grammar, or any out-of-grammar utterance. In addition, if this application is able to detect when the caller requests an unsupported name or city, it can respond appropriately (for example, "Sorry, but there's no office in that location").

Also note the pattern for the introduction, which has the general sequence Welcome–Purpose–More Info, followed by an Explicit Prompt. Usually the explicit prompt at the end of the introduction is enough to guide the caller to provide an appropriate response; however, if the caller input after that explicit prompt is an out-of-grammar utterance, something from the Help grammar, or a silence timeout, then the system falls through a sequence of prompts designed to provide progressively greater assistance. The beauty of this system is that the caller never hears the same prompt twice, so the system appears to be responsive to the caller and always moving the dialog forward; this help style promotes efficient use by most callers and avoids the negative social consequences of saying the same error message repeatedly. If the caller still fails to provide an appropriate response, the sequence of prompts ultimately "bails out" (typically by transferring to a human agent or apologizing and exiting).

In some specific cases it might be necessary to treat help, nomatch, and noinput events differently, but those are the exceptions. Usability testing in our lab has shown, however, that callers who say "Help" once do not expect the system to provide different messages or reprompts as a result of saying "Help" again. For this reason, it is a good practice to modify the duration of noinput timeouts for that dialog step from the standard 7 seconds to 3 seconds following an explicit request for help. This increases the likelihood that the caller who has requested help will hear the help message appropriate to the situation.

Bailing Out

Deciding how many attempts to allow the caller before bailing out (leaving the speech application) is a judgment call. You do not want to bail out after the first unsuccessful attempt, or you will lose callers too quickly; conversely, you do not want to require too many attempts, or you run the risk of frustrating callers who are experiencing serious problems with the system. For most applications, if callers are experiencing serious problems with the system it is better to get them out of the system quickly and, if possible, give them another means to access the desired information (say, by pointing them to a visual Web site, switching to a touchtone mode, or transferring them to an agent). In general, two or three progressively more directed prompts (including a presentation of the global navigation commands) should suffice to help callers recover from noinput events, nomatch events, or explicit requests for help (Rolandi, 2004a). For example:

System: Transfer $500 from checking to savings? <2 sec pause> Please say Yes, No, or Repeat.

Caller: (nomatch or noinput)

System: If you want to transfer $500 from checking to savings, say Yes (or press 1). Otherwise, say No (or press 2).

Caller: (nomatch or noinput)

System: At any time you can say Repeat, Help, Go Back, Start Over, Transfer to Agent, or Exit. <2000 ms pause> To continue, please say Yes, No, or Repeat.

Caller: (nomatch or noinput)

System: Please hold for the next available agent.

Summary

Before starting detailed dialog design, it's necessary to make a number of high-level design decisions. For most IVR applications, enable speech (as opposed to hotword) barge-in. To as great an extent as possible, use professionally recorded audio for system output. If using TTS, minimize the juxtapositions between TTS and recorded speech.

Pay attention to an application's persona, but not too much. For most service-based applications, the persona should match that of highly skilled call center agents—assuming the caller is busy, being efficient when communication is good, helpful when progress is slower, polite, rarely apologizing, and never blaming the caller. Emphasize a clear, simple, and efficient design based on user goals, avoiding expensive investment in biographical persona details that do not contribute to detailed design.

Match the complexity of the requisite speech technologies to the needs of the application's users. If simple speech recognition will suffice, then use the simpler technologies. If callers will significantly benefit from menu flattening or form flattening, then consider making the investment in more complex technologies.

Favor concise over verbose initial prompting. For most IVR applications, plan for a touchtone fallback to allow callers who experience difficulty using speech to continue self-service. Consider enabling a small number of global navigation functions including, as appropriate for the application, Repeat, Help, Go Back, Start Over, Transfer to Agent, and Exit.

Applications for enterprises seeking to provide a high level of customer service will include the ability to transfer to human agents. The design of support for such transfers will differ for enterprises with call centers that are always open and those that are not. Do not make an up-front announcement of agent availability, but be prepared to deal appropriately with early requests for an agent. Design choices for transfer requests include:

When the call center is open:

Transferring the caller as requested

Negotiating with the caller to continue using the IVR

Offering a call back option

When the call center is closed:

Telling the caller about the call center hours

Offering a call back option

For callers who need help, provide self-revealing contextual help rather than a help mode. When necessary, but only when necessary, treat noinput, nomatch, and requests for help differently. When there is evidence that a caller is having difficulty using a speech application, help him or her get to an alternative channel, for example, switching to a touchtone version of the application or transferring to an agent. Do not move callers out of an application after a single incident, but do not wait for more than two or three consecutive incidents in a dialog step before bailing out.

References

Ahlén, S., Kaiser, L., & Olvera, E. (2004). Are you listening to your Spanish speakers? *Speech Technology, 9*(4), 10–15.

Attwater, D. (2008). *Speech and touch-tone in harmony [PowerPoint Slides]*. Paper presented at SpeechTek 2008. New York, NY: SpeechTek.

Bailly, G. (2002). Close shadowing natural versus synthetic speech. *International Journal of Speech Technology, 6*, 11–19.

Balentine, B. (2006). The power of the pause. In W. Meisel (Ed.), *VUI visions: Expert views on effective voice user interface design* (pp. 89–91). Victoria, Canada: TMA Associates.

Balentine, B. (2007). *It's better to be a good machine than a bad person*. Annapolis, MD: ICMI Press.

Balentine, B. (2010). Next-generation IVR avoids first-generation user interface mistakes. In W. Meisel (Ed.), *Speech in the user interface: Lessons from experience* (pp. 71–74). Victoria, Canada: TMA Associates.

Balentine, B., Ayer, C. M., Miller, C. L., & Scott, B. L. (1997). Debouncing the speech button: A sliding capture window device for synchronizing turn-taking. *International Journal of Speech Technology, 2*, 7–19.

Balentine, B., & Morgan, D. P. (2001). *How to build a speech recognition application: A style guide for telephony dialogues* (2nd ed.). San Ramon, CA: EIG Press.

Blattner, M., Sumikawa, D., & Greenberg, R. (1989). Earcons and icons: Their structure and common design principles. *Human-Computer Interaction, 4*, 11–44.

Boyce, S. J. (2008). User interface design for natural language systems: From research to reality. In D. Gardner-Bonneau & H. E. Blanchard (Eds.), *Human factors and voice interactive systems* (2nd ed.) (pp. 43–80). New York, NY: Springer.

Boretz, A. (2009). VUI standards: The great debate. *Speech Technology, 14*(8), 14–19.

Brems, D. J., Rabin, M. D., & Waggett, J. L. (1995). Using natural language conventions in the user interface design of automatic speech recognition systems. *Human Factors, 37*(2), 265–282.

Byrne, B. (2003). "Conversational" isn't always what you think it is. *Speech Technology, 8*(4), 16–19.

Byrne, B. (2004). In the studio: Setting high standards for prerecorded audio. *Speech Technology, 9*(2), 46–48.

Callejas, Z., & López-Cózar, R. (2008). Relations between de-facto criteria in the evaluation of a spoken dialogue system. *Speech Communication, 50*, 646–665.

Cohen, M. H., Giangola, J. P., & Balogh, J. (2004). *Voice user interface design.* Boston, MA: Addison-Wesley.

Commarford, P. M., Lewis, J. R., Al-Awar Smither, J., & Gentzler, M. D. (2008). A comparison of broad versus deep auditory menu structures. *Human Factors, 50*(1), 77–89.

Crystal, T. H., & House, A. S. (1990). Articulation rate and the duration of syllables and stress groups in connected speech. *Journal of the Acoustical Society of America, 88*, 101–112.

Dahl, D. (2006). Point/counter point on personas. *Speech Technology, 11*(1), 18–21.

Damper, R. I., & Soonklang, T. (2007). Subjective evaluation of techniques for proper name pronunciation. *IEEE Transactions on Audio, Speech, and Language Processing, 15*(8), 2213–2221.

Dialogues Spotlight Research Team. (2000). *The grouping of numbers for automated telephone services.* Edinburgh, UK: Centre for Communication Interface Research.

Dobroth, K. (2000). Beyond natural: Adding appeal to speech recognition applications. *Speech Technology, 5*(1), 20–23.

Dougherty, M. (2010). What's universally available, but rarely used? In W. Meisel (Ed.), *Speech in the user interface: Lessons from experience* (pp. 117–120). Victoria, Canada: TMA Associates.

Francis, A. L., & Nusbaum, H. C. (1999). Evaluating the quality of synthetic speech. In D. Gardner-Bonneau (Ed.), *Human factors and voice interactive systems* (pp. 63–97). Boston, MA: Kluwer Academic.

Fröhlich, P. (2005). Dealing with system response times in interactive speech applications. In *Proceedings of CHI 2005* (pp. 1379–1382). Portland, OR: ACM.

Gaver, W. (1986). Auditory icons: Using sound in computer interfaces. *Human Computer Interaction, 2*, 167–177.

Gong, L., & Lai, J. (2003). To mix or not to mix synthetic speech and human speech? Contrasting impact on judge-rated task performance versus self-rated performance and attitudinal responses. *International Journal of Speech Technology, 6*, 123–131.

Gong, L., Nass, C., Simard, C., & Takhteyev, Y. (2001). When non-human is better than semi-human: Consistency in speech interfaces. In M. J. Smith, G. Salvendy, D. Harris, & R. J. Koubek (Eds.), *Usability evaluation and interface design: Cognitive engineering, intelligent agents and virtual reality* (pp. 390–394). Mahwah, NJ: Lawrence Erlbaum.

Gorin, A. L., Riccardi, G., & Wright, J. H. (1997). How may I help you? *Speech Communication, 23*, 113–127.

Hafner, K. (2004, Sept. 9). A voice with personality, just trying to help. *The New York Times*. Retrieved from www.nytimes.com/2004/09/09/technology/circuits/09emil.html

Hambleton, M. (2000). Directing the dialog: The art of IVR. *Speech Technology, 5*(1), 24–27.

Henton, C. (2003). The name game: Pronunciation puzzles for TTS. *Speech Technology, 8*(5), 32–35.

Hura, S. L. (2008). What counts as VUI? *Speech Technology, 13*(9), 7.

Karray, L., & Martin, A. (2003). Toward improving speech detection robustness for speech recognition in adverse conditions. *Speech Communication, 40,* 261–276.

Klie, L. (2007). It's a persona, not a personality. *Speech Technology, 12*(5), 22–26.

Kotelly, B. (2003). *The art and business of speech recognition: Creating the noble voice.* Boston, MA: Pearson Education.

Kuo, H. J., Siohan, O., & Olive, J. P. (2003). Advances in natural language call routing. *Bell Labs Technical Journal, 7*(4), 155–170.

Lai, J., Karat, C.-M., & Yankelovich, N. (2008). Conversational speech interfaces and technology. In A. Sears & J. A. Jacko (Eds.) *The human-computer interaction handbook: Fundamentals, evolving technologies, and emerging applications* (pp. 381–391). New York, NY: Lawrence Erlbaum.

Larson, J. A. (2005). Ten guidelines for designing a successful voice user interface. *Speech Technology, 10*(1), 51–53.

Lee, C.-H., Carpenter, B., Chou, W., Chu-Carroll, J., Reichl, W., Saad, A., & Zhou, Q. (2000). On natural language call routing. *Speech Communication, 31,* 309–320.

Lewis, J. R., Commarford, P. M., & Kotan, C. (2006). Web-based comparison of two styles of auditory presentation: All TTS versus rapidly mixed TTS and recordings. In *Proceedings of the Human Factors and Ergonomics Society 50th annual meeting* (pp. 723–727). Santa Monica, CA: Human Factors and Ergonomics Society.

Lewis, J. R., Simone, J. E., & Bogacz, M. (2000). *Designing common functions for speech-only user interfaces: Rationales, sample dialogs, potential uses for event counting, and sample grammars* (Tech. Rep. 29.3287, available at drjim.0catch.com/always-ral.pdf). Raleigh, NC: IBM Corp.

Lines, L., & Hone, K. S. (2002). Older adults' evaluations of speech output. In *Proceedings of the Fifth International ACM Conference on Assistive Technologies* (pp. 170–177). Edinburgh, Scotland: ACM.

Lombard, E. (1911). Le signe de l'elevation de la voix. *Annales des maladies de l'oreille et du larynx, 37,* 101–199.

Massaro, D. (1975). Preperceptual images, processing time, and perceptual units in speech perception. In D. Massaro (Ed.), *Understanding language: An information-processing analysis of speech perception, reading, and psycholinguistics* (pp. 125–150). New York, NY: Academic Press.

McInnes, F., Attwater, D., Edgington, M. D., Schmidt, M. S., & Jack, M. A. (1999). User attitudes to concatenated natural speech and text-to-speech synthesis in an automated information service. In *Proceedings of Eurospeech99* (pp. 831–834). Budapest, Hungary: ESCA.

McKellin, W. H., Shahin, K., Hodgson, M., Jamieson, J., & Pichora-Fuller, K. (2007). Pragmatics of conversation and communication in noisy settings. *Journal of Pragmatics, 39,* 2159–2184.

Polkosky, M. D. (2001). *User preference for system processing tones* (Tech. Rep. 29.3436). Raleigh, NC: IBM.

Polkosky, M. D. (2005a). *Toward a social-cognitive psychology of speech technology: Affective responses to speech-based e-service*. Unpublished doctoral dissertation. University of South Florida.

Polkosky, M. D. (2005b). What is speech usability, anyway? *Speech Technology, 10*(9), 22–25.

Polkosky, M. D., & Lewis, J. R. (2001). *The function of nonspeech audio in speech recognition applications: A review of the literature* (Tech. Rep. 29.3405, available at drjim.0catch.com/audlitrev-ral.pdf). West Palm Beach, FL: IBM.

Roberts, L. A., Silver, E. M., & Rankin, L. L. (2004). Selecting a voice persona. In *Proceedings of the Human Factors and Ergonomics Society 48th annual meeting* (pp. 712–716). Santa Monica: HFES.

Rolandi, W. (2003). The common causes of VUI infirmities. *Speech Technology, 8*(6), 29.

Rolandi, W. (2004a). Improving customer service with speech. *Speech Technology, 9*(5), 14.

Rolandi, W. (2004b). Rolandi's razor. *Speech Technology, 9*(4), 39.

Rolandi, W. (2007). The persona craze nears an end. *Speech Technology, 12*(5), 9.

Rosenfeld, R., Olsen, D., & Rudnicky, A. (2001). Universal speech interfaces. *Interactions, 8*(6), 34–44.

Sauro, J., & Lewis, J. R. (2005). Estimating completion rates from small samples using binomial confidence intervals: Comparisons and recommendations. In *Proceedings of the Human Factors and Ergonomics Society 49th annual meeting* (pp. 2100–2104). Santa Monica, CA: Human Factors and Ergonomics Society.

Schegloff, E. A. (2000). Overlapping talk and the organization of turn-taking for conversation. *Language in Society, 29*, 1–63.

Shriver, S., & Rosenfeld, R. (2002). Keywords for a universal speech interface. In *Proceedings of CHI 2002* (pp. 726–727). Minneapolis, MN: ACM.

Spiegel, M. F. (1997). Advanced database preprocessing and preparations that enable telecommunication services based on speech synthesis. *Speech Communication, 23*, 51–62.

Spiegel, M. F. (2003a). Proper name pronunciations for speech technology applications. *International Journal of Speech Technology, 6*, 419–427.

Spiegel, M. F. (2003b). The difficulties with names: Overcoming barriers to personal voice services. *Speech Technology, 8*(3), 12–15.

Stern, S. E., Mullennix, J. W., & Yaroslavsky, I. (2006). Persuasion and social perception of human vs. synthetic voice across person as source and computer as source conditions. *International Journal of Human-Computer Studies, 64*, 43–52.

Stevens, C., Lees, N., Vonwiller, J., & Burnham, D. (2005). On-line experimental methods to evaluate text-to-speech (TTS) synthesis: Effects of voice gender and signal quality on intelligibility, naturalness and preference. *Computer Speech and Language, 19*, 129–146.

Suhm, B. (2004). Lessons learned from deploying. *Speech Technology, 9*(6), 10–12.

Suhm, B. (2008). IVR usability engineering using guidelines and analyses of end-to-end calls. In D. Gardner-Bonneau & H. E. Blanchard (Eds.), *Human factors and voice interactive systems* (2nd ed.) (pp. 1–41). New York, NY: Springer.

Suhm, B., Bers, J., McCarthy, D., Freeman, B., Getty, D., Godfrey, K., & Peterson, P. (2002). A comparative study of speech in the call center: Natural language call routing vs. touch-tone menus. In *Proceedings of CHI 2002* (pp. 283–290). Minneapolis, MN: ACM.

Tomko, S., Harris, T. K., Toth, A., Sanders, J., Rudnicky, A., & Rosenfeld, R. (2005). Towards efficient human machine speech communication: The speech graffiti project. *ACM Transactions on Speech and Language Processing, 2*(1), 1–27.

Turunen, M., Hakulinen, J., & Kainulainen, A. (2006). Evaluation of a spoken dialogue system with usability tests and long-term pilot studies: Similarities and differences. In *Proceedings of the 9th International Conference on Spoken Language Processing* (pp. 1057–1060). Pittsburgh, PA: ICSLP.

Wang, H., & Lewis, J. R. (2001). Intelligibility and acceptability of short phrases generated by embedded text-to-speech engines. In *Proceedings of HCI International 2001: Usability evaluation and interface design* (pp. 144–148). Mahwah, NJ: Lawrence Erlbaum.

Weegels, M. F. (2000). Users' conceptions of voice-operated information services. *International Journal of Speech Technology, 3,* 75–82.

Yang, F., & Heeman, P. A. (2010). Initiative conflicts in task-oriented dialogue. *Computer Speech and Language, 24,* 175–189.

8

Getting Specific: Low-Level Design Decisions

After deciding the high-level properties of a system, it's time to work out the low-level issues, especially regarding specific interaction styles and prompts. This chapter covers:

- Creating introductions
- Avoiding poor practices in introductions
- Getting the right timing
- Designing dialogs
- Constructing appropriate menus and prompts
- Recovering from errors
- Confirming caller input

Creating Introductions

Because 100% of calls to an IVR go through its introductory messages it is very important to craft an introduction that accomplishes its goals as concisely as possible (Rolandi, 2004, 2007a; Suhm, 2008). Every unnecessary syllable in an introduction to an often-used IVR can cause expense for the enterprise and wastes the caller's time. The cost savings per second differs from application to application, but Yudkowsky (2008) reported a savings for AT&T of $1 million per year for each second of extraneous speech removed from a (presumably) very high-volume IVR.

Figure 8.1 illustrates common elements of introductions. As the figure shows, the required elements are an introductory welcome message (first) and an initial prompt (last), with intervening optional elements of language selection, statement of purpose, listing of major global navigation commands, statement of ability to interrupt at any time, and transition from optional elements to the initial prompt ("Let's get started").

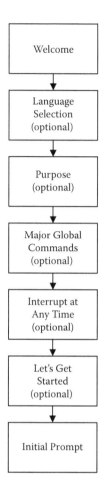

FIGURE 8.1
Common elements of IVR introductions.

Welcome (Required)

The welcome prompt should be short and simple, making it clear that the caller is talking to a machine. Human agents rarely begin their conversations with the word "Welcome," so that works well as the first word in IVR introductions. For example:

System: Welcome to AutoRez car reservations.

If the application requires echo calibration for barge-in to work properly (the IVR software engineers should know this), then it might be necessary to disable barge-in during the introduction. If barge-in is disabled during this initial welcome message, it is important to trim all silence from the end of this audio segment, then, after enabling barge-in, play a short amount of silence to separate the initial welcome message from the following audio segment. Otherwise, silence at the end of the initial audio can act as a turntaking cue, inviting callers to barge in, but will cause usability problems because the application has not yet enabled barge-in. If a sufficiently high percentage of users will be repeat callers, consider pausing for about 1 second after the initial welcome to make it more comfortable for experienced callers (who know what to say) to barge in (Balentine & Morgan, 2001). Then continue playing the remaining introductory prompts.

Language Selection (Optional)

More and more IVRs serve a multilingual caller base (Ahlén, Kaiser, & Olvera, 2004; Klie, 2010). In the United States, the most common languages are English and Spanish; in Canada, English and French. From the perspective of providing the shortest possible path to the initial prompt, the best solution for serving a multilingual caller base is to provide a different phone number for each language (Balentine, 2007). Many enterprises, however, prefer to have a single phone number, in which case it is necessary to provide some means for language selection.

The best solution to the problem of providing concise language selection is to make the nonprimary language option a choice on the initial menu (usually placed on the 9 key), rather than creating a separate language selection dialog step. For example, use:

> System: Para Español, oprima nueve. <Continue on to next prompt, allowing callers to press 9 at any time during the next prompt to switch to Spanish>

rather than:

> System: Para Español, oprima uno. <Timeout to give Spanish-speaking callers time to press 1>

Using "uno" (1) for the language selection requires establishing a timeout during which this choice will be active, ending when the initial menu becomes active. These types of timeouts, which place an extreme restriction on selection time, are difficult to design and can lead to selection errors. The timeouts also increase the mean call duration for primary-language callers, who must wait for the timeout to end before the rest of the introductory messages can begin to play in the primary language.

With the preferred approach ("Para Español, oprima nueve."), there is no need for the timeout to give speakers of the nonprimary language time to respond before the IVR continues in the primary language. The option to press 9 for Spanish is available until the caller makes a selection from the initial menu. Using a touchtone key rather than a speech prompt for language selection avoids a number of issues associated with multilingual language selection via speech, is very concise, and is extendible to more than one alternate language, within reason.

Purpose (Optional)

If necessary, state the purpose of the system. For example:

System: With AutoRez, you can rent a car at any major airport in the US.

There is no need to include a statement of purpose if the welcome prompt adequately conveys the system's purpose. Also, do not include a statement of purpose unless necessary because it lengthens the introduction.

Short List of Global Commands (Optional)

Next, carefully consider if it's reasonable to tell the caller about key global commands. First-time callers will rarely remember the commands, but this presentation might be helpful for repeat callers.

If it is reasonable, then list just two or three. For example:

System: You can always say Repeat, Help, or Go Back.

I've used this type of message in a number of applications and it does no obvious harm, but I don't know of any research on its effectiveness. Because playing this type of message, no matter how concise, delays the caller from hearing the initial prompt, it's probably best to deploy the application without it. Just in case, though, it might be a good idea to include it in the recording manifest so it will be available if post-deployment analysis of the application indicates a need for it.

Interrupt at Any Time (Optional)

When systems have barge-in enabled, some introductions include a message to let callers know they can interrupt the application. For example:

System: And you can interrupt me at any time.

For most user populations, it isn't necessary to provide this type of message because callers will naturally interrupt the application when it pauses. Heins, Franzke, Durian, and Bayya (1997) studied the barge-in behavior of two groups of participants who used a speech application with barge-in enabled. One group of participants received explicit instruction about the barge-in capability of the system; the other did not. There was no difference in the actual barge-in behavior of the two groups, with both groups showing a strong tendency to barge in during pauses in system speech.

Despite the evidence against it, I have had clients who have insisted on including this type of message. As with any message of dubious value, if pushed to include it, craft a succinct version. Using our primary vendor at current prices, adding the example version of the message to an existing recording manifest would add about $6.00 to the cost. Thus it is reasonable to have the recording available to deploy if post-deployment measures or future experimentation indicated a need for this type of message.

Let's Get Started (Optional)

To provide a concise transition from the introductory messages to the initial prompt, it can be helpful to include a short phrase such as:

System: Let's get started!

The "Let's get started" phrase, however, should be considered only if the introduction has included one or more of the optional messages. Otherwise, it would serve no purpose and should not be part of the introduction.

Initial Prompt (Required)

Finally, present the caller with the first prompt requiring user input. For example:

System: You can make, review, change, or cancel reservations. What would you like to do?

Putting It All Together

Table 8.1 shows the complete sample introduction with all optional elements, including syllable counts and estimated durations (assuming about 200 ms per syllable, Crystal & House, 1990; Massaro, 1975). Counting syllables manually is more time-consuming than using automatic word-counting facilities, but a syllable count provides a better source for estimating and comparing estimates of prompt durations. Don't make too much of small differences in syllable counts, but counting syllables is an inexpensive empirical method for comparing the efficiency of alternative prompts and messages. If the introduction included all the elements shown in Table 8.1, the amount of time callers would have to wait from the beginning of the welcome message to the end of the initial prompt would be about 17.4 seconds. If the application required a language prompt, then eliminating the rest of the optional messages would result in an introduction lasting 8.2 seconds (2.2 + 2.0 + 4.0)—less than half the duration of the "complete" introduction, saving about 9.2 seconds per

TABLE 8.1

Sample Introduction

Element	Sample Text	Syllables	Estimated Duration (s)
Welcome	Welcome to AutoRez car reservations.	11	2.2
Language selection	Para Español, oprima nueve.	10	2.0
Purpose	With AutoRez, you can rent a car at any major airport in the United States.	20	4.0
Global commands	You can always say Repeat, Help, or Go Back.	11	2.2
Interrupt at any time	And you can interrupt me at any time.	11	2.2
Let's get started	Let's get started!	4	0.8
Initial prompt	You can make, review, change, or cancel a reservation. What would you like to do?	20	4.0

call. Clearly the more efficient design is the one that gets callers to the initial prompt as quickly as possible.

Avoiding Poor Practices in Introductions

Balentine (2007) contains an invaluable essay titled "Avoiding Bad Practice," which addresses many other common messages associated with IVR introductions—messages generally acknowledged in the speech user interface (SUI) IVR design community to be poor practice, including:

- Web deflection messages
- Sales pitches
- Prompts for touchtone versus speech
- "Your call is important to us"
- "Please listen carefully as our options have changed"
- "This call may be monitored or recorded"

Web Deflection Messages

Some enterprises desperately want to get callers to abandon the call and use the Internet instead because the cost per transaction is so much lower for the Web (Rolandi, 2005, 2007a). The rationale (from Balentine, 2007) for avoiding Web deflection messages in the introduction is that the goal of successful Web deflection requires the completion of four successful sub-goals.

- Remind the caller that an enterprise Web site exists.
- Get the caller to believe that the Web experience will be superior to the IVR.
- Get the caller to know the right Web address after hearing it once.
- Get the caller to hang up the phone and use the Web instead.

However:

- It's the 21st century. Callers already know that enterprise Web sites exist.
- The caller has already chosen the phone as the preferred channel of communication—if the caller believed the Web experience would be superior or more appropriate at the moment, he or she would already be on the Web.
- Unless the Web address is very simple, it can be difficult to accomplish the third subgoal. For a company such as Prudential, the Web address is very simple—www.prudential.com. For a company such as Philips, it's more complicated because there are two common spellings for the name, with the more common spelling having two rather than one "l" ("Phillips").
- There is no evidence that callers frequently (if ever) hang up and go to the Web after hearing this kind of message in an IVR's introduction.

When presented in an IVR's introduction, Web deflection messages are typically just time wasters, more likely to irritate than to please callers. Inappropriate placement of Web deflection messages has adverse effects on usability, corporate image, and the cost per call. "For all those callers who don't need or can't use the web site address, the company spends an extra 6¢ in telephone charges on every call. After 40 million calls each year, that begins to add up" (Kotelly, 2006, p. 62). Web messages have their place in IVRs (for example, in after-hours messages), but not in the introduction.

Sales Pitches

Avoid sales pitches in the introduction of an IVR. Users call IVRs to accomplish specific tasks, and hearing a sales pitch is rarely (if ever) the caller's goal. Like Web deflection messages, sales pitches can have a place in the IVR (for example, presenting an up-selling or other personalized message when callers indicate they are ready to end the call), but not in the introduction. "Poorly placed advertisements that inappropriately take up the caller's time or offer directions that are unlikely to be followed, are a hit against the brand—unless the brand stands for slow, thoughtless service. This behavior frustrates the caller and wastes money" (Kotelly, 2006, p. 62).

Prompts for Touchtone versus Speech

Because some callers might prefer touchtone to speech, it can be tempting to prompt for this choice in the introduction. However, as Attwater (2008) has pointed out, callers care more about receiving effective service and care less about the modality of the user interface than its immediate usability. Balentine (2007) further argued that prompting for the choice has several drawbacks:

- It forces all callers (most of whom don't care) to listen to the choice, only to serve those who have a preference
- It delays the caller's first task-oriented interaction

For these reasons, it's better to assume that speech interaction will work, switching to touchtone automatically following the detection of sufficient evidence (consecutive noinput or nomatch events) of caller difficulty with speech. If it isn't working, then make the necessary changes to get it to work—testing for recognition accuracy, tuning recognition accuracy, testing for usability, etc. For the input of numeric data (account numbers, PINs, SSN, etc.), it's a common and effective practice to use a "say or enter" prompt so callers know that touchtone is available.

Your Call Is Important to Us

Regarding messages such as "We value your call" or "Your call is important to us," Balentine (2007, p. 363) noted, "The enterprise that values a customer's call will service the call quickly. It is counterproductive to waste precious time on declaring value when the declaration itself defeats the value." Rolandi (2005, p. 22) stated, "The practice is widespread, but the absurdity of the situation should be obvious if one adopts the perspective of the caller." The general consensus is that this type of message wastes time at best, and at worst, is simultaneously annoying.

Please Listen Carefully as Our Options Have Changed

Balentine (2007) argues against putting this message in the introduction, even if the options have recently changed. Callers believe that they are listening carefully, so that part of the message ("Please listen carefully …") is annoying. Designers put the message into the opening menu because they've made a change to the menu, and want to alert repeat callers (experts) to the change. Experts, however, often skip past this message unless it's forced, but forcing callers to hear a message annoys experts even more than novices. As a result, the message fails to serve the target caller and penalizes other callers. Furthermore, many callers who listen to the message don't believe it because these types of messages have a tendency to hang around long after the menu changes have become ancient history (and callers know this).

If required to include this message, make it as concise as possible and avoid suggesting that callers listen carefully (for example, "Please note that our options have changed," or possibly, "Our options changed December 3rd, 2010"), and consider playing it only for callers who have not heard it yet. After it's been deployed for a month or so, get rid of it, preferably through the automatic triggering of a programmed setting.

This Call May Be Monitored or Recorded

Don't play a "calls may be monitored or recorded" message in the introduction unless:

- There is a legal obligation to play the message
- The IVR is actually recording

If including this type of message, make it reasonably concise, for example:

System: Your call may be monitored or recorded.

As concise as this message is, it is still 11 syllables in length, with an estimated duration of about 2.2 seconds. If a cost-benefit analysis warrants the investment (additional code and testing), design the application to play this message when the IVR is recording and to disable it when not recording.

Getting the Right Timing

Timing is important in developing a usable, conversational dialog rhythm in SUI design (Balentine, 2006). Despite its importance, it often tends to be underspecified in detailed dialog specifications. The consequence of this underspecification is that someone other than the designer decides, either deliberately or accidentally, what the duration of pauses will be in the implemented application. Research in the appropriate duration of pauses in SUI design is ongoing, but there is sufficient published research to provide some specific guidance to designers.

Specific timing decisions depend on the context in which the pause appears. For example:

- The noinput timeout value for dialog steps
- The length of pauses between options in a menu
- The pause between initial prompts and prompt extensions
- The time it takes the system to respond to a caller utterance

Research on Turntaking Behaviors in Speech Recognition IVRs

Research on the timing of turntaking in dialogs is of particular importance to SUI design. Summarizing from Chapter 3 ("Timing and Turntaking"), research in human–human turntaking behavior has shown:

- Pauses associated with conversational turntaking in a face-to-face setting rarely last more than 1 second (Clark, 1996; Wilson & Zimmerman, 1986)
- When a participant in a telephone conversation pauses longer than 1 second and the other participant does not take the turn, it is common for the first participant to interpret this as indicative of a problem (Roberts, Francis, & Morgan, 2006)
- The strongest verbal cue that a speaker wants to yield the floor (in other words, has completed a turn and is ready for someone else to begin talking) is for that speaker to stop talking (Johnstone, Berry, Nguyen, & Asper, 1994)
- The pause duration for dialog turns in service-based telephone conversations was 426 ms, with a 95% confidence interval ranging from 264–588 ms, an estimated 95th percentile of 1010 ms, and an estimated 99th percentile of 1303 ms (Beattie & Barnard, 1979)

These findings from human–human communication suggest that pauses in an IVR's system speech of about a quarter of a second (250 ms) should not typically trigger turntaking behaviors, whereas system pauses longer than 1300 ms are very likely to do so. Findings from research on turntaking behaviors in speech recognition IVRs are generally consistent with those predictions.

Heins et al. (1997) studied a call-blocking service that used a speech recognition application with barge-in enabled. The participants included eleven men and nine women. Caller tasks included recording greetings, enabling and disabling greetings, and modifying call-screening features. Even after 16 repeated exposures to prompts such as, "To confirm, say OK. To cancel, say cancel," participants did not barge into the prompt, but waited until it finished, speaking at a location that syntax, prosody, silence, and experience pointed to as an appropriate place in the dialog to take the floor. The researchers noted (p. 162) that "when listeners self-select, they will wait until they have all necessary information to take their turn, but they tend to use the next possible syntactic boundary to break in." There were no significant differences detected in caller performance or attitude as a function of age, gender, or familiarity with high technology.

Margulies (2005) reported a series of studies of caller turntaking behavior when using speech recognition IVRs. In one study 100 survey respondents provided information about their experiences with speech recognition IVRs; specifically, how they knew when it was their turn to speak. The most commonly reported cue was a pause in the dialog (41.6%), followed by the syntax of the prompt (26.2%), then the inflection at the end of the prompt (21.0%), and finally, earcons (11.2%).

In another study, Margulies (2005) analyzed videos of several dozen participants interacting with speech recognition IVRs. The major causes of failures in interaction were system floorholding and rigid timeouts. System floorholding, caused by extraneous messages, long and mostly unnecessary instructions, and pauses too short to encourage turntaking on the part of the caller, accounted for 62.5% of failures. Margulies' definition of "rigid timeouts"—the cause of 18.75% of the dialog failures—was "when the machine seemingly yields a turn but then continues with instructions or a repeat of the declarative or interrogatory prompt—coincident with the subject either preparing to respond or in the act of responding" (p. 7).

The use of the term "rigid" for this phenomenon is somewhat misleading. Strictly speaking, all timeouts are rigid (inflexible) because they are either specified in the application code or are included in audio recordings. Rather than being due to rigidity, the likely underlying problem is that the timeout durations that occurred in these dialog steps were either unspecified or incorrectly specified. Participants in the study expressed a great deal of frustration when the system would "step on" them. Margulies concluded that the underlying cause of both types of turntaking errors (floorholding and incorrect timeout specification) was "a stingy application of silence" (p. 7), and recommended that SUI designers should:

- Use silence as the basic building block for turntaking cues.
- Yield the floor often (in other words use silence liberally and provide sufficient time for callers to comfortably begin speaking).
- Avoid lengthy instructions.

Setting the Noinput Timeout Value

The noinput timeout value is the amount of silence the system must detect before triggering a noinput event. For VoiceXML, the default noinput timeout value is 7 seconds. In general, this default seems to work well, although shortening it to 5 seconds also seems to work well for most populations and standard dialog steps (Yuschik, 2008). Based on analyses of about 100 recorded interactions with phone-based systems from a series of usability tests, Margulies (2005) recommended that noinput timeout values be no less than 3 seconds. For applications designed primarily for special populations such as non-native speakers or older adults (Dulude, 2002), it's reasonable to increase this to a larger value (say, around 10 seconds). The same applies for specific dialog steps in which the caller must perform an action such as getting a credit card out of a purse or looking up an account number. In those cases, temporarily adjust the timeout value to match the estimated time needed for the action.

Timing Pauses between Menu Items

For much of the history of speech recognition IVRs, speech menus played a dominant role. Even for more modern SUI designs that strive to be conversational, it is occasionally necessary to present a list of options to callers for their selection. To enhance the usability of speech menus consider using the following timing guidelines:

Use a 750-1000 ms pause between items when there are more than three items (for example, "Please select checking, <750 ms> savings, <750 ms> money market, <750 ms> or loans"). When there are only two or three items, do not introduce any exaggerated pauses. Speak the phrase as a normal sentence (for example, "Please select checking or savings.").

If leading with a directive phrase such as "select" or "please select" to enhance the conversational tone, avoid placing detectable pauses between "select" and the first option.

For touchtone prompts or touchtone-style speech prompts (for example, "For checking, press 1. For savings, 2. For money market, 3. For loans, 4. <2 sec pause> To speak with a representative, press 0.") use the following timing guidelines:

- Use a 500-750 ms pause between items.
- Use a 250 ms pause before "press <x>" or "<x>," where "<x>" is the number.
- There should be no detectable pause after "for," "to," or "press."

Research on working memory limitations and turntaking protocols provides the primary foundation for these recommendations, along with considerations for communicative efficiency. When there are only two or three concise options, speaking them as a normal sentence allows their production without significant inter-option pausing and without placing undue stress on the caller's working memory. When there are more options, it is necessary to change the mode of production. One way to accomplish this, as described

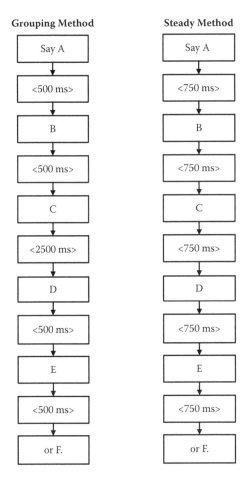

FIGURE 8.2
Two timing schemes for voice menus.

above, is to place a significant pause between each option to allow callers time to mentally process the options as they play so there is no need to memorize each option, with the inter-option time also acting as a turntaking cue that invites callers to speak after hearing the desired option (Balentine, 1999; Balentine, Ayer, Miller, & Scott, 1997; Commarford, Lewis, Al-Awar Smither, & Gentzler, 2008).

Yuschik (2008) published a different way to accomplish similar goals. Rather than providing a uniform pause between each option, based on a series of international usability studies, he recommended playing options in groups of up to three items, with short pauses (500 ms) between items and longer pauses (2500 ms) between the groups. Figure 8.2 depicts a six-option menu presented using (1) Yuschik's grouping method and (2) an alternative steady method, similar to the method used in Commarford et al. (2008).

Both methods control the amount of information that callers must retain in memory before having a period of silence in which to process what they have heard, staying within the limits of human working memory. The two methods have similar amounts of total silence, 4500 ms for grouping, 3750 ms for steady as illustrated in Figure 8.2 and as typically deployed (4500 ms for steady if using 900 ms rather than 750 ms inter-item pauses). Both methods use an "or" before the last option in the list as a syntactic turntaking cue.

There is no published research directly comparing these two methods, but I currently recommend the steady method over the group method because:

- The pacing of the presentation of information in the steady method is more consistent, and therefore should be more predictable. In particular, callers could interpret the long silence following the presentation of the third option in the group method as a strong turntaking cue that would lead them to respond before hearing the remaining options.

- Although both methods control the amount of information that callers must hold in memory to accomplish the selection task, the memory demands of the steady method should be more uniform during the selection process than the group method, which very likely produces peaks of demand at the end of the presentation of each group of options.

- The organization of menu options is unlikely to cluster neatly into groups of three. The steady method should match up better than the grouping method with a wide variety of menus.

- The steady method is more consistent with the research that recommends frequent pauses of sufficient length to encourage callers to take the conversational floor (Heins et al., 1997; Margulies, 2005). The 500 ms inter-item pause of the grouping method is close to the average pause reported by Beattie & Barnard (1979), so it is probably a less reliable turntaking cue than the slightly longer pauses of 750–1000 ms used in the steady method.

Note that the differential demands on working memory between the two methods are not caused by the memorization of all options, but instead are due to the working memory processes required to process and evaluate the items as they come in. If items come in too rapidly, then the caller will need to hold the next item in memory while still working on the previous item and remembering the best item so far—a potentially cascading effect that could overload working memory. On the other hand, given the delivery of items at a steady pace with sufficient time between (750–1000 ms), callers will have the time they need to complete processing and evaluating an option and will be ready to attend to the next one if they haven't yet made a decision, while still comfortably remembering the best option so far (see "Optimal Menu Length" later in this chapter, especially Figure 8.5).

Timing Pauses between Prompts and Prompt Extensions

Sometimes designers provide prompts followed almost immediately by prompt extensions (Weegels, 2000). For example:

System: Was that Delta 1 2 3 4? <500 ms pause> Please say yes or no.

or:

System: What type of call would you like to make? <500 ms pause> Please select collect, calling card, or person to person.

or:

System: <Plays phone mail message>. Would you like to Delete, Reply to, or Forward this message? <500 ms pause> At any time you can say Repeat, Help, Go Back, or Start Over.

A 500 ms pause, as shown in these examples, is consistent with the length of pauses that speakers typically leave at the ends of sentences when speaking more than one sentence, is close to the average turntaking pause reported by Beattie & Barnard (1979), and is a very common pause duration to encounter in deployed speech recognition IVRs. However, a 500 ms pause between a prompt and its extension is problematic due to its turntaking ambiguity—the longer the pause, the less ambiguous it is as a turntaking cue. For some callers, the 500 ms pause provides just enough time to mentally process what the system just said and to begin to respond with an answer to the question. But just as some callers begin to speak, the system starts to speak the prompt extension, which "steps on" the caller, interfering with the ongoing mental processing, causing significant frustration (Margulies, 2005) and, for some callers, triggering a stuttering effect. Like the stuttering effect sometimes triggered by hotword-type barge-in (see Chapter 7, "Lombard Speech and the Stuttering Effect"), this can lead to a cascade of errors after the system fails to recognize the stuttered utterance.

I once had the opportunity to review videos of a usability test of a speech recognition IVR designed to support a state's family services. The speech recognition IVR had numerous instances of prompts and extensions separated by roughly 500 ms pauses. Table 8.2 shows the effect of this on one of the participants as she tried to get information about an upcoming court date appointment for a juvenile court hearing. The transcript in the table uses asterisks to indicate overlapping speech from one row to the next.

TABLE 8.2

Example of Turntaking Problems Due to Inappropriate Timing

Row	Speaker	Transcript of Spoken Dialog
1	System	Main menu. <100 ms> Say Payments, <250 ms> Appointments, <150 ms> PIN Change, <288 ms> General Info, <168 ms> or More Options.
2	Participant	Appointments.
3	System	Which would you like? <100 ms> Intake Interviews, <100 ms> Genetic Testing, <100 ms> Court Date Info or Upcoming Appointments? <450 ms> *Or say, "It's none of these."*
4	Participant	*Court date info,* <272 ms> Court date *info.*
5	System	*You have* <1500 ms> Say Repeat That, <315 ms>, Next Appointment, <272 ms>, Main Menu, or Goodbye. <608 ms> To speak to someone, say Agent.
6	Participant	<2300 ms> Court date *appointment.*
7	System	*Sorry.* <1500 ms> Sorry, I didn't get that. <200 ms> Say Repeat That or press 1, <408 ms> Next app …
8	Participant	<Presses 1>
9	System	You have an interview coming up on September 5th at 9:20 a.m. <10 ms> Say Repeat That, <315 ms>, Next Appointment, <272 ms>, Main Menu, or Goodbye.
10	Participant	<272 ms> Main menu.
11	System	Main menu. <100 ms> Say Payments, <250 ms> Appointments, <150 ms> PIN Change, <288 ms> General Info, <168 ms> or More Options.
12	Participant	Appointments.
13	System	<1675 ms> Which would you like? <100 ms> Intake Interviews, <100 ms> Genetic Testing, <100 ms> Court Date Info or Upcoming Appointments? <450 ms> *Or say,* "It's none of these."
14	Participant	*Court date*, <280 ms> Court date *info*
15	System	*Sorry,* <1600 ms> I need to know what type of appointment you're interested in so that I can transfer you to the right agent for assistance. <712 ms> Please say Intake Interviews, <232 ms> Genetic Testing, <376 ms> Court Date Info, *or Upcoming Appointments*.
16	Participant	*Court date info.*
17	System	<1650 ms> You have a juvenile court hearing coming up on August 6th at 9:25 a.m. <10 ms> Say Repeat That, <315 ms>, Next Appointment, <272 ms>, Main Menu, or Goodbye.
18	Participant	Goodbye.
19	System	Did you say …
20	Participant	<Hung up>

The timing problems started in Row 3. After presenting a list of options, the system waited 450 ms, then spoke a prompt extension ("Or say, 'It's none of these.'"). This prompt extension stepped on the participant as she said, "Court date info." The system actually understood this utterance, and started to say "You have" (Row 5), just as the participant, who thought the system had not heard her the first time, repeated "Court date info" (Row 4). This utterance barged into the system just as it was about to play the target information (Row 5).

Unfortunately, the phrase "Court date info" was no longer in grammar, so the system played a list of the available options (end of Row 5). The participant, still trying to get court date information but having forgotten the exact wording, said "Court date appointment" (Row 6). The system response to this out-of-grammar utterance (Row 7) was to apologize for not "getting that," and then to offer a menu of choices with speech and touchtone options. The participant pressed "1" to repeat (Row 8), then heard appointment information, but not the desired court date information (Row 9). She elected to start over by saying "Main menu" (Row 10). The system went back to the main menu (Row 11), and at exactly the same point as in Row 3 the timing led her to stutter again on "Court date info," leading to similar confusion. Finally, in Row 17 she got the desired information.

During the post-task interview, the participant stated, "This time she [the system] like, kind of cut you off too fast. … It's like there's not a long enough pause in there for you to say what you [want to say]. … So I think maybe if there's a bit of a longer pause in there to give you time to talk … it would be a lot better."

If 500 ms is a poor choice for the pause between a prompt and its extension, then what is a better choice? The available research suggests that the duration should be at least 1500 ms, and in some cases a bit more. Brems, Rabin, and Waggett (1995) conducted a series of studies to understand how much of a pause should be between an open-ended prompt (such as, "What type of call would you like to make?") and a complete listing of the valid call types ("Please say collect, calling card, third number, person to person, or operator."). Their final recommendation was to separate the initial prompt and its extension by 1500 ms. When playing example prompts after an initial, open-ended prompt in a natural language (NL) call routing application, Sheeder and Balogh (2003) recommended a pause of 2500 ms between an initial nondirective prompt and its extension ("How can I help you? <2500 ms> You can ask me about things like 'minutes used,' 'automatic payments,' and 'calling plans.' So, what can I help you with?").

The same principle applies to the presentation of global commands following a set of end-of-message options (options presented at a terminal point—a point at which the caller might have finished getting the desired information from the IVR). Commarford and Lewis (2005) analyzed usability testing videos of six callers at task-terminal points in the completion of tasks with two different speech recognition IVRs. The goal of the analysis was to determine the optimal pause between initial presentation of a menu at a task-terminal point and the presentation of global navigation commands as a prompt extension (see Chapter 7, "Using the Global Commands"). There was no consistent effect of menu length (from 2 to 10 options in the initial speech menu) on the time for callers to respond. As shown in Figure 8.3, there were some interesting differences in the distributions of caller response latencies as a function of whether the terminal menu included or did not include

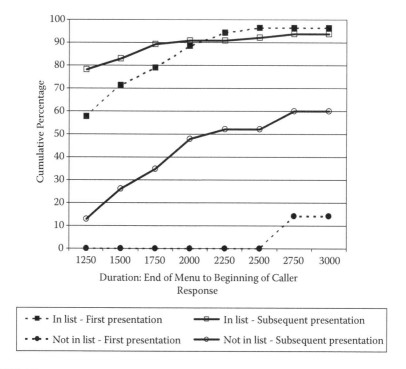

FIGURE 8.3
Percentage of caller responses starting within 1250–3000 ms.

the target option for the task and whether it was the first or a subsequent presentation of the menu to the caller.

When a desirable option was in the selection list for the initial presentation of the menu callers started a selection within 2500 ms 96.2% of the time (within 1250 ms 57.7% of the time), and made no selections in the time period ranging from 2500–3000 ms. When callers had previously heard the menu items, they typically responded a little more quickly, with 78.1% of responses starting within the first 1250 ms. In the seven clear cases in which no desirable options appeared in a menu presented for the first time users never began a response within the first 2500 ms. However, when subsequently presented with the menu (including those instances in which the first presentation did include an option that met user needs), users tried an unhelpful command 52.2% of the time within 2500 ms. These data also show that users typically responded more quickly to menus with which they were familiar, as indicated by the percentage of responses given in less than 1250, 1500, and 1750 ms. The data for "Not in list—Subsequent presentation" indicate that these types of prompt extensions should begin playing before 2500 ms have elapsed after the end of the base prompt. The data for "In list—First presentation" and "In list—Subsequent presentation" hit about 90% at a duration of 2000 ms and

94% at 2500 ms. Thus a 2000 ms pause seems reasonable to use between prompts and prompt extensions.

The following example shows a 2000 ms pause between the initial set of task-terminal options and the global navigation commands.

> Caller: (Listens to a baseball news story.)
>
> System: Select next, repeat, or previous.
>
> Caller: Next.
>
> System: (Plays next baseball news story.)
>
> System: Select next, repeat, or previous. <2000 ms> At any time you can say Repeat, Help, Go Back, Start Over, or Exit.
>
> Caller: Start Over.
>
> System: I've got new baseball, football, and basketball stories. What would you like to hear?

If an extension has been added in an attempt to fix a usability problem with a prompt, then the intervening pause should be shorter than 500 ms—say, closer to 150–200 ms—to ensure that the extension starts playing before the caller begins to speak ("Was that Delta 1 2 3 4? <150 ms> Please say yes or no."). Alternatively, it might be better (less disruptive of mental processing of the primary prompt) to rearrange the elements to provide stronger and more consistent timing and syntactic turntaking cues, as in:

> System: Yes or no, was that Delta 1 2 3 4?

Adding this type of phrase before the actual prompt seems to have a very directive tone, however, which is out of character with the typical service provider persona. Before adding an extension or lead-in to a prompt that isn't working well, spend some time trying to improve the prompt itself. When including a prompt like this in an application, make sure that the tone of the recording is as appropriate as possible. There is a clear need for additional research in this area, in particular, studies of the appropriate timing as a function of open-ended versus more directive prompts and as a function of whether the prompt is for information the caller has likely memorized or must think about before responding.

Managing Processing Time

A common industry goal is for 95% of system responses to occur within two seconds of the end of a caller's input (Balentine & Morgan, 2001; Fried & Edmondson, 2006). Some callers might interpret a long processing delay as an indication that the system did not hear or did not accept the most recent input. This can cause callers to repeat what they just said, which can lead to misrecognitions and accidentally stopping a prompt that has just begun to play. SUI designers have no control over the existence of processing delays, but there are techniques for managing the caller experience.

In an unpublished IBM usability study of a speech-recognition IVR, we placed an artificial delay of 8 seconds following a prompt in the prototype under test, with just the two-word phrase "Please wait" playing after the caller's input and before the delay. All callers stayed silent during the delay, which suggests that this simple instruction worked to manage processing times of up to 8 seconds.

Fröhlich (2005) inserted artificial delays from 2 to 16 seconds in duration into a speech recognition prototype of mobile voice services such as news, traffic, cinema, and weather information. In one study, he observed caller behaviors (32 participants) during unannounced waiting periods. In another test, he had 27 participants rate the waiting times according to "pleasantness of waiting time" and "necessity of a waiting cue." About half of the participants received a spoken confirmation of input ("OK"); the other half did not. Those who received input confirmation also experienced tasks in which they heard an announcement ("The requested information is being processed. Please hold the line.") then music. In a follow-on study 13 participants experienced waiting times of 16 and 32 seconds, starting with an announcement of the expected wait time and with the remainder of the wait time filled with music.

In Fröhlich's (2005) first study, participants did not demonstrate overt signs of impatience during an unannounced delay until the duration of the silence was 12 seconds. Confirming user input with "OK" significantly improved pleasantness ratings. Through delays of 4 seconds, ratings of pleasantness were about equal for silent delays and delays filled with a waiting cue (such as simple musical loop—also see Chapter 7, "Deciding Whether to Use Audio Formatting"). For delays greater than 4 seconds, silence became less and less pleasant, and ratings of the need for a waiting cue increased. For the longer delays (16 and 32 seconds), the announcement of the expected waiting time followed by music received very positive pleasantness and appropriateness ratings (also see Chapter 4, "Waiting for Service").

Boyce (2008) reported a study of caller tolerance of system delay. During a call, callers experienced a delay from 2 to 8 seconds in duration, filled with either silence or a musical sound effect. There were 45 participants in the study, with each participant experiencing, in random order, every possible combination of delay and filler. After each call, participants rated the delay as acceptable or unacceptable. Callers found delays of 2–3 seconds acceptable, with or without the sound effect. Only when the delay was longer than 6 seconds did callers rate the sound effect as significantly more acceptable than silence.

A summary of these research results appears in Table 8.3. The research does not support hard and fast rules here, but it does suggest general strategies. When expecting short delays (durations consistent with common

TABLE 8.3

Summary of Research Results for Managing Processing Time

Expected Delay	Acknowledge Input	Announce Waiting Time	Play Processing Tone
2–3 seconds	Optional	No	Optional
4–8 seconds	Yes	No	Yes
9–15 seconds	Yes	Optional	Yes
> 16 seconds	Yes	Yes	Yes

industry goals for system response time), it makes no sense to announce the expected waiting time because there's hardly enough time to make the announcement. Apparently there is no harm in a concise acknowledgment of input ("OK" or "Please wait") or processing tones for short delays, but neither is there any apparent benefit. For longer expected delays, however, the benefits of acknowledging input and playing a processing tone increase. For expected delays longer than 15 seconds, there is plenty of time to play a waiting time announcement and such an announcement seems to have a clear benefit in improving the caller experience associated with the delay. The research summarized here does not address the required accuracy of such announcements, but it is reasonable to assume, until such evidence is available, that wildly inaccurate estimates could do more harm than good.

Designing Dialogs

IVR dialogs begin with a prompt. Prompts can be directive (explicit) or nondirective (implicit) (Yankelovich, 1996). Whether directive or nondirective, designers should word options from the caller's point of view, using terms familiar to the caller and avoiding excessive complexity (Suhm, 2008). Although anthropomorphism in IVRs has been controversial among researchers in human–computer interaction (Gardner-Bonneau, 1999), it is currently common to script applications using the personal pronouns "I" and "me," with some evidence that callers prefer these applications even though they do not explicitly notice the use of the personal pronouns (Boyce, 2008). Even so, it is important to keep the user in focus when designing dialogs, avoiding applications that are excessively self-referential (Balentine, 2007).

When carefully written and contextually appropriate, applications can include both directive and nondirective prompting. Experiments conducted by Enterprise Integration Group (2000) showed that "an application can successfully employ both NL and directed dialogue design types. Users can move between the two approaches, in a single application, without either confusion or failure to accomplish tasks. Since each approach has its unique value, designers can safely build an application using both. In particular, when the directed dialogue is kept simple, users are seldom aware of it so the application appears to users to be seamless" (p. 22).

Writing Directive Prompts

When writing directive prompts, designers must decide whether to write highly directive prompts or more conversational prompts. The attraction of highly directive prompts is that they don't leave much to the imagination, and tend to result in highly predictable responses because callers tend to use the same words they heard in the prompts (Zoltan-Ford, 1991). The attraction of more conversational directive prompts is the desire to match caller expectations with regard to customer service behavior and speech characteristics, while still avoiding excessive verbosity (Byrne, 2003; Polkosky, 2005a, 2005b). For example, the following is a highly directive prompt:

> System: Select Checking, Savings, or Money Market.

This is a more conversational version:

> System: Do you want to work with your checking, savings, or money market account?

The more conversational version (18 syllables) is longer than the highly directive version (11 syllables), but not excessively verbose. Caller responses to the more conversational version will be more variable than responses to the highly directive version, so it's necessary to have more robust grammars when using more conversational prompts (Bush & Guerra, 2006). Despite some reservations that caller responses to these types of conversational prompts might include just "yes" or "no," experience shows that callers usually respond as expected (Joe, 2007), although their responses will sometimes include "yes," as in "Yes, checking please." Foley and Springer (2010) noted that callers sometimes anticipate the desired information for prompts that the designer intended to produce a simple yes or no response. For example:

> System: Is there a three-digit code on the back of your credit card?
> Caller: Three five seven.
> System: Please say Yes, No, or Repeat. Is there a three-digit code on the back of your credit card?

This prompting and error recovery was part of an application that I worked on many years ago, and fortunately we caught the design problem early in usability testing. I still remember the first participant who encountered it rolling his eyes, responding "Yes," and then continuing with the dialog. We changed the grammar for that dialog step to accept "Yes" or "No" and several variants, including the ability to recognize three digits either with or without variants of "Yes," and had no further problems with that step.

Making Nondirective Prompts Usable

The initial prompts for NL understanding applications (those that perform NL routing with or accept multiple tokens in the caller's response) are usually nondirective, for example, "How may I help you?" or "What would you like to do?" Despite their ability to promote more efficient interaction with an application when callers know what to say, a major usability issue with nondirective prompts is that many callers do not quickly work out how to respond. For example, Walker, Fromer, Di Fabbrizio, Mestel, & Hindle (1998) studied user preference for two spoken language interfaces to e-mail, one directive and the other nondirective. The initial directive prompt was "Hi, Elvis here. You have five new and zero unread messages in your inbox. Say Read, or Summarize, or say Help for more options." The nondirective version was "Hi, Elvis here. I've got your mail." Even though the nondirective version was more efficient as measured by the mean number of turns or elapsed time to complete a set of tasks, participants significantly preferred the directive version. The problem with the nondirective version seemed to be confusion about the available options and poor speech recognition for the callers' unconstrained utterances. McInnes, Nairn, Attwater, Edgington, and Jack (1999) reported a similar finding, with more successful caller responses to "Which service do you require?" and "Please say 'help' or the name of the service you require" than "How can I help you?"

One way to improve the usability of nondirective prompts is to provide example prompting (Balentine & Morgan, 2001; Enterprise Integration Group, 2000; Knott, Bushey, & Martin, 2004; Sheeder & Balogh, 2003; Suhm, Bers, McCarthy, Freeman, Getty, Godfrey, & Peterson, 2002; Williams & Witt, 2004), in the spirit of what Karsenty (2002) described as shifting from an invisible to a transparent SUI design philosophy.

Design decisions associated with example prompting are:

- How many examples to provide
- Exactly what example or examples to use
- Where to place the examples—before, directly after, or following a pause after the nondirective prompt
- If after a pause, then how long to wait before playing the example prompt(s)

I know of no research comparing the number of examples to provide. It is rare in practice to provide more than three, and some designers argue for just one to avoid the appearance of a set menu of options (Barkin, 2009; Joe, 2007). It is important to provide at least one, but designers should take care

TABLE 8.4

Initial Prompting in Sheeder and Balogh (2003)

Condition	Initial Prompting
Keyword Following	Welcome to Clarion Wireless Customer Service. How can I help you? <2500 ms> You can ask me about things like Minutes Used, Automatic Payments, and Calling Plans. So, what can I help you with?
Natural Following	Welcome to Clarion Wireless Customer Service. How can I help you? <2500 ms> You can ask me things like "How many minutes have I used?" and "I'd like to set up automatic payments." So, what can I help you with?
Keyword Preceding	Welcome to Clarion Wireless Customer Service. You can ask me about things like Minutes Used, Automatic Payments, and Calling Plans. So, how can I help you with your account?
Natural Preceding	Welcome to Clarion Wireless Customer Service. You can ask me things like "How many minutes have I used" and "I'd like to set up automatic payments." So, how can I help you with your account?

if providing more than one to ensure that each additional example provides value to the caller.

Sheeder and Balogh (2003) described an experiment in which they manipulated two variables—the positioning of examples (preceding the open-ended prompt or following the prompt after a 2.5 second noinput timeout) and the type of example (keyword list or more natural expression). There were 18 participants for each of the four possible conditions (preceding keyword, preceding natural, following keyword, following natural) for a total of 72 participants. Participants completed three telephone account management tasks (activate service, bill reprint, and change plan) with their assigned condition. Table 8.4 shows examples of the initial prompting for these four conditions. All conditions included a 2.5 second pause followed by "For more assistance, just say 'help'" at the end.

Sheeder and Balogh (2003) concluded that the natural preceding condition had the best outcomes, with a significantly higher successful task completion rate. In the following conditions, 87% of the callers spoke during the 2500 ms pause, so only 13% heard the examples. Sheeder and Balogh pointed out that it is possible that a following strategy might be the better approach for applications in which most callers are expected to be frequent, experienced users of the application because it would be more efficient. For applications in which most callers will use the application infrequently, there appeared to be an advantage in providing examples before the open-ended prompt.

Based on data from six initial prompting styles (directed versus open prompts, preceded by hello, hello plus an earcon, and hello plus an earcon plus a named persona plus a recently changed notice), Williams and Witt (2004) concluded that the style that produced the most routable utterances was a directed prompt preceded by hello plus an earcon plus a named persona plus a recently changed notice ("[earcon] Hello, welcome to Acme.

My name is Johnson, your virtual assistant. Please take note, this service has recently changed."). It is important, however, to take this as a provisional recommendation since the difference in percentage of routable calls between that condition and a simpler condition (hello plus an earcon—"[earcon] Hello, welcome to Acme") was not statistically significant ($\chi^2(1) = 2.4$, p = .12), and the generally accepted leading practice is to avoid playing unnecessary information in the introduction (e.g., avoid playing recently changed messages or naming the application's persona). Because the open prompt did not include examples in this experiment (as in Walker et al., 1998 and McInnes et al., 1999), it is unsurprising that the directed dialog produced a higher percentage of routable utterances.

At least for the application designs investigated in their first experiment, the directed designs of Williams and Witt (2004) were more successful than the open designs, possibly because callers did not have a strong mental model of the task. When the application design was closer to the designs investigated by Sheeder and Balogh (2003) the routability percentages were higher, but it didn't seem to matter if the examples appeared before or after their open question ("What would you like to do?"). As Williams and Witt acknowledge, this could be due to differences in the applications studied or the types of participants (usability test participants in Sheeder and Balogh; real callers in Williams and Witt).

The focus of research by Blanchard and Stewart (2004) was on how to reprompt an open-ended prompt under two situations (a caller asks for a human; a caller's initial response is too vague to classify). They created reprompts for these two situations and found that the following reprompts were effective for the application they were studying:

- After receiving a request for an agent: "OK. In order to direct your call please tell me if you need to refill an existing medication, find out the status of your prescription order, or anything else you want to speak to the customer service representative about."
- After receiving a vague request about an order: "I'm sorry, do you need to refill an existing medication or get status on an order you've already sent in? Please tell me how I may help you.
- After receiving any other vague request: "OK. What's your question?"

Note that while these specific prompts were found to be effective in getting callers to provide more specific responses, Blanchard and Stewart (2004) did not explicitly test their style of priming the response (a fairly natural-sounding but short list of keywords) against the more commonly practiced style of example prompting ("For example, you might say 'refill an existing medication,' 'tell me the status of a prescription order,'" etc.).

Knott, Bushey, and Martin (2004) conducted experiments similar to that of Sheeder and Balogh (2003), but used a deployed system to capture initial

TABLE 8.5

Initial Prompting in Knott, Bushey, and Martin (2004)

Condition	Initial Prompting
Experiment 1: No Example	"Welcome to SBC. I'm here to help you reach the right place. First, please tell me the purpose of your call, and then press the pound key."
Experiment 1: Example	"Welcome to SBC. I'm here to help you reach the right place, so please tell me why you're calling today, and then press the pound key. You can say things like, 'What's my account balance?' or, 'I'd like to get some rates for long distance.' So, how can I help you?"
Experiment 2: No Example	"Welcome to SBC. Hi, I'm Bill and I'm here to help you reach the right place. Please tell me why you're calling today, and then press the pound key. So, how can I help you?"
Experiment 2: Example	"Welcome to SBC. Hi, I'm Bill and I'm here to help you reach the right place. Please tell me the purpose of your call, and then press the pound key. You can say things like, 'What's my account balance?' or, 'I'd like to get some rates for long distance.'"

caller responses to two open-ended prompts—one that started with examples and one that never presented any examples. In Experiment 1 the participants were 4058 callers from California and Nevada; in Experiment 2 there were 4479 callers from Illinois, Indiana, Michigan, Ohio, and Wisconsin. For the prompts used in Experiment 1, the application did not refer to itself by a name, but in Experiment 2 it did ("Hi, I'm Bill"). Table 8.5 shows the initial prompts used in the experiments. Note the unusual instruction to "press the pound key." The researchers (p. 737) stated, "This instruction was necessary for the speech data capture application, but would not normally be included in a natural language speech prompt." Because this instruction was the same in all four experimental conditions it should not have strongly influenced the outcomes.

Providing examples up front improved the routability of the caller's initial utterance—a finding consistent with that of Sheeder and Balogh (2003). Knott, Bushey, and Martin (2004) also found, however, that providing examples up front increased the likelihood that the caller would request an agent rather than making an attempt to provide a routable input (from 4.5–9.4% in Experiment 1; from 5.4–7.9% in Experiment 2). They also found that when the application did not refer to itself by name, there was no significant difference in the wordiness of caller responses as a function of presenting examples up front, but when the system did refer to itself by name, caller responses were more verbose in the no examples condition. As with the results of Sheeder and Balogh (2003), the findings regarding the placement of examples are very likely limited to applications in which callers are infrequent users of the application rather than frequent users. In addition to the primary findings of the experiments, callers had a strong tendency to mimic the examples presented in the prompts, emphasizing the importance of selecting examples that the system can recognize with high accuracy and that are requests that callers frequently make.

When using nondirective prompts with a pause before example prompts, the silence timeout that precedes example prompting should be shorter than the standard default timeout of 7 seconds—typically about 2–3 seconds—to serve the dual purpose of providing a sufficient window for the more expert caller to compose and begin speaking a multiple-token command but not making the less expert caller wait too long to get some help. For applications that can accept multiple tokens, research (Balentine & Morgan, 2001; Enterprise Integration Group, 2000) has suggested that examples for these types of prompts should include valid multiple-token commands. Callers are often able to take a multiple-token example and replace the sample information with their own information. Only if these fail to lead the caller to produce a valid input should the system switch to a one-token-at-a-time directed dialog. The following illustrates the successful use of an example prompt:

System: What are your travel needs?

Caller: <nomatch or silent for 3 seconds>

System: For example, you might say "I want to go from New York to Orlando on December first."

Caller: I want to go from Chicago to Los Angeles on March fifteenth.

And this illustrates unsuccessful use of example prompts:

System: What are your travel needs?

Caller: <nomatch or silent for 3 seconds>

System: For example, you might say "I want to go from New York to Orlando on December first."

Caller: <nomatch or silent for 3 seconds>

System: You might also try "Book a flight on Delta from Chicago to Miami on November fifteenth."

Caller: <nomatch or silent for 7 seconds>

System: Let's try a different way. What's your departure date?

Multi-token example prompts should use concrete examples rather than the names of variables. For example, use "For example, you might say, 'I want to go from New York to Orlando on December first,'" rather than "For example, you might say, 'I want to go from departure city to arrival city on travel date.'" According to Balentine and Morgan (2001), prompts that use variable names are "confusing and ineffective" (p. 207).

Hura (2006) described a slightly different strategy for getting callers to be successful with multi-token recognition in a banking application for which most users would be frequent callers. If a caller provided a single token at the initial nondirective prompt, her application accepted that information and moved on to collect the remaining data one prompt at a time. "We reasoned that there was no particular value in making the user aware of multislot recognition initially, and using directed dialog for the first transfer has the added benefit of providing implicit instruction on what the required slots are and how to fill them. We chose to present our example as a hint after the user has successfully completed a transfer (Here's a hint. Next time you can tell me the whole transfer at once, like this: 'Transfer $100 from my checking account to savings.')" (Hura, 2006, pp. 80–81). In this application, the hints were dynamically generated from the data provided by the users, who reacted positively to the design and successfully used the multi-token functionality in future calls. Table 8.6 lists recommendations for the design of nondirective prompts.

Mixed-Mode Prompting

An important part of the craft of SUI design of IVRs is the appropriate use of touchtone (Attwater, 2008; Balentine, 2010; Kaiser, 2006). Initial prompts should generally not attempt to mention both speech and touchtone (Balentine & Morgan, 2001). The application interface is simpler when prompts (at least initial prompts) focus primarily on speech or are neutral with respect to the mode ("Next, what's your PIN?"). This doesn't mean that it is necessary to disable touchtone input during the initial prompt—just don't specifically prompt for it.

An exception to this is prompting for numeric strings such as ZIP codes, which commonly start with "Say or enter ...," a concise and well-understood way to let callers know that they have the choice to speak or key in the information. Suhm (2008) reported that analyses of end-to-end calls for applications using "say or enter" prompts for the entry of numeric strings have consistently shown that 60% of callers choose to use the keypad when offered the alternative.

Attwater (2008), after analyzing data over several years from large sample live traffic tests of IVRs in the United States and Europe, found that callers strongly prefer touchtone for numeric data strings, with 90% of callers using touchtone in response to the prompts "Say or enter your bank account number" and "Say or enter your telephone number." For seminumeric data, most callers (60%) used speech in response prompts such as "Say or enter your date of birth", although a substantial minority did use touchtone. When offered a dual-mode menu prompt ("Say A or press 1, B or press 2, or C or press 3"), the response pattern was 65% touchtone and 35% speech. Attwater concluded that IVRs should support both touchtone and speech where possible.

Although not as highly recommended as starting with speech only, if there is a strong desire to offer touchtone during initial menu presentations, then

TABLE 8.6

Key Recommendations for the Design of Nondirective Prompts

Recommendation	Rationale
1. Support nondirective prompts with one to three naturally phrased example prompts.	Example prompts teach callers how to provide valid responses to the prompt and help to establish context. If using multiple examples write them in a way that does not imply that they are a menu of options and ensure each example has its purpose.
2. For applications that will have many repeat callers, place examples after the nondirective prompt; for those with few repeat callers, place examples before the nondirective prompt.	Repeat callers will generally know what to say so will benefit less from example prompts. For prompts that will accept multiple tokens consider providing a multi-token hint after the caller has successfully provided all the necessary data. Providing example prompts before the nondirective prompt increases the likelihood of a caller providing a valid initial response, but slows down the interaction for frequent callers.
3. If placing examples after the nondirective prompt, start with a 3 second pause before the presentation of the examples.	For these types of prompts, a pause of 2–3 seconds seems to offer enough time for callers who have an idea about what to say to respond to the prompt without getting stepped on, but provides example prompts quickly enough for callers who will benefit from hearing them. Be sure to design the application so it is easy to adjust the duration of this pause in a tuning step.
4. Choose example prompts that convey phrases that the application can recognize with high accuracy and that are requests that callers frequently make.	There's no point in offering examples that have low relevance to callers and are hard for the application to recognize.
5. Consider using a nondirective prompt that asks for brief input ("briefly," "in a few words") if a more commonly used but less imperative version (such as, "How may I help you?") isn't working well.	There have been claims that nondirective prompts such as "In a few words, please tell me why you're calling" or "Please tell me, briefly, the reason for your call today" lead to more tractable utterances from callers (Suhm et al., 2002). However, the published evidence is weak and this type of initial prompting is longer and less service oriented than a "How may I help you" prompt. Boyce (2008) reported a slight advantage for the use of the word "briefly" when performing an open-ended reprompt, but that appears to be less effective than example prompting.

use maximally concise addition of the touchtone information to the initial prompt, for example:

System: Select Checking (or press 1), Savings (2), or Money Market (3).

For this approach to work as effectively as possible, the touchtone numbers need to play in close proximity to their associated functions, and at about half the volume of the surrounding audio.

"Press or Say <X>" Prompting

Developers who convert applications from touchtone to speech sometimes choose a "Press or say <x>" user interface:

> System: For checking, press or say 1. For savings, press or say 2. For money market, press or say 3.

Although this is a common approach to mixed-mode prompting, it is far from leading practice. "Press or say <x>" user interfaces are better than plain touchtone user interfaces because callers can use them without having to move the handset away from their faces to find the right number to press. The main problem with "Press or say" user interfaces is that they inherit the well-known weaknesses of touchtone user interfaces and fail to take advantage of the strengths of well-designed SUIs. Some of these well-known weaknesses are:

- "Press or say" prompts tend to be wordier than better-designed speech prompts—longer to play and thus slower
  - For example, consider, "For checking, press or say 1. For savings, press or say 2. For money market, press or say 3." (23 syllables).
  - An appropriate speech prompt is, "Select checking, savings, or money market," which is 11 syllables shorter—about a 50% (2.2 second) reduction.
- Even more important, "Press or say" user interfaces require callers to remember two things about the choices that are presented in the menu
  - The content of the choice
  - The number associated with the choice

For callers who have lower than average memory span (limited capability to hold information in short-term memory, which becomes more of a problem with aging populations), it can be much more difficult to use the system because these types of user interfaces require at least twice as much memory capacity as a preferred-practice speech menu, in which each options' content is the same as what the caller should say.

Balentine and Morgan (2001) specifically address the use of "Press or say" user interfaces in Section 7.2.1.1 (p. 190):

> 7.2.1.1 Avoid "Press or Say" if Possible: Asking the user to speak a digit for menu selections or other non-numeric data—simply to emulate the touch-tone keypad—is extremely awkward. Although speech recognition technologies of several years ago were limited to such vocabularies, this is no longer the case.

So why are "Press or say <x>" user interfaces so prevalent? For one thing, it's the easiest way to speech enable an existing touchtone application. Also, imagine that you're someone who has a lot of touchtone design experience, and now you have the task of designing a speech application. Unless you've made an effort to investigate preferred practices in SUI design, then this (a "Press or say <x>" user interface) is the way you're likely to think about structuring the interface. As mentioned before, however, even though it has some advantages over plain touchtone, it is far from leading practice in SUI design. In a recent Web search, I found numerous examples of applications using a "Press or say <x>" user interface, but no examples of SUI designers promoting the style (Lewis, 2007).

Also, some stakeholders believe that a "Press or say <x>" user interface is more robust than other SUI styles. Part of this belief is based on the fact that a "Press or say <x>" SUI uses smaller grammars than other SUI styles and will, as a result, have higher recognition accuracy. It is true that the grammars are smaller, but this doesn't necessarily translate to higher recognition accuracy when contrasted against a well-designed alternative. Part of leading practice in designing a directed dialog is to ensure that each option has enough acoustic information for the application to recognize it accurately and that all options are phonetically distinct—advantages that a simple digits grammar cannot match even in high-noise settings.

Constructing Appropriate Menus and Prompts

A characteristic of many voice applications is that they have little or no external documentation. Often, these applications must support both novice and expert callers. Part of the challenge of designing a good voice application is providing just enough information at just the right time. In general, don't force callers to hear more than they need to hear, and don't require them to say more than they need to say. The topics in this section are:

- Research on optimal menu length
- Grouping menu items, prompts, and other information
- Minimizing error declarations
- Avoiding touchtone-style prompts
- Choosing a complex alternative
- Crafting effective prompts
- Managing transitions between task steps and non-task steps
- Stepping stones and safety nets

Optimal Menu Length

Menus have played a significant role in IVR design for decades, starting with touchtone applications. With the improvements in speech recognition and development of natural language technologies, some SUI designers and researchers have argued strongly against continuing to use menus in speech recognition applications (Gardner-Bonneau, 1999; Harris, 2005). Others have pointed out that auditory menus, or at least, selection from a spoken list of options, occurs during normal customer service dialogs (Balentine, 1999; Rolandi, 2007b), for example, "Would you like that small, medium, or large?" or "Which dressing? We've got Thousand Island, French, honey mustard, oil and vinegar, raspberry vinaigrette, or blue cheese." For conciseness of expression and consistency with industry terminology, I'll refer to these as auditory menus. For the foreseeable future, and possibly forever, crafting effective auditory menus will be part of practical SUI design.

Two key characteristics of an auditory menu (actually, any menu) are its breadth and depth (Commarford et al., 2008; Virzi & Huitema, 1997). For a given number of options presented in an auditory menu, which is the better strategy?

- Fewer menus with more options per menu (a broader, flatter design)
- More menus with fewer options per menu (a deeper design)

Most early auditory menu design guidelines suggested a limit of four to five options per menu (Gardner-Bonneau, 1992; Gould, Boies, Levy, Richards, & Schoonard, 1987; Marics & Engelbeck, 1997; Schumacher, Hardzinski, & Schwartz, 1995; Voice Messaging User Interface Forum, 1990), a recommendation carried forward into current SUI design by many practitioners and researchers (Balentine & Morgan, 2001; Cohen, Giangola, & Balogh, 2004; Suhm, 2008; Wilkie, McInnes, Jack, & Littlewood, 2007). The cited rationale for this limit is generally Miller's (1956) famous paper about the magic number 7 ± 2, part of which described experiments demonstrating that people have trouble remembering more than about 7 (plus or minus 2) items at a time due to limits in human working memory (Baddeley & Hitch, 1974). For a given total number of options, restricting the number of options per menu necessarily leads to a deeper rather than a broader menu structure, as shown in Figure 8.4.

Since 1997, research in both touchtone and speech menu design has shown that the guideline to limit menus to four or five options is more of a misguideline (Huguenard, Lurch, Junker, Patz, & Kass, 1997; Virzi & Huitema, 1997). Using research in and theories of human working memory Huguenard et al. created a cognitive model phone-based interaction (PBI USER). Their architecture suggested that "it is not the number of options per menu level that determines the magnitude of WM [working memory] load, but rather the amount of processing and storage required to evaluate the 'goodness' of each individual option. Thus WM load should rise, peak, and

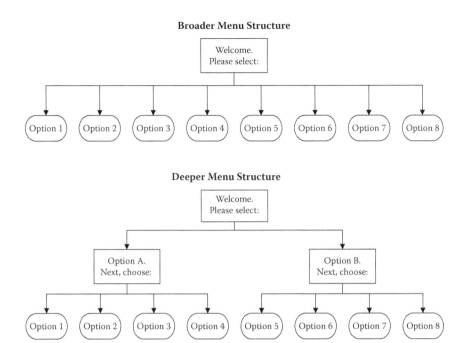

FIGURE 8.4
Broader and deeper menu structures for design with eight options.

fall during each option evaluation, and it is the height of each peak relative to total WM capacity that determines the probability of information loss" (p. 75). In an experiment with 87 participants, Huguenard et al. compared performance with two navigational menu structures that each contained 81 terminal nodes, one broader (two levels, 9 × 9) and one deeper (four levels, 3 × 3 × 3 × 3), and found that participants in the deep condition experienced more navigational errors than those in the broad condition. Their primary practical conclusion was that deep menu hierarchies do not appear to reduce working memory error rates in phone-based interaction.

Virzi and Huitema (1997) studied a commonly used method to constrain the number of options presented in auditory menus—the presentation of up to four options, then a fifth "More Options" choice that provides a list of the remaining options. Note that this is similar to Yuschik's (2008) grouping method (see Figure 8.2), except callers must explicitly select the "More Options" choice to hear the additional options, rather than having the additional items automatically presented after a 2500 ms pause. In their experiment, Virzi and Huitema had 24 participants interact with broad and deep versions of four different touchtone IVR menu applications, each of which had eight target options. They manipulated familiarity with the applications by having participants interact with one application 16 times, a second application 8 times, a third application 4 times, and the fourth application 1 time. The placement of target options varied from trial to trial, with two targets from the first set of four options and two targets from the second set. Thus their design had three independent variables: menu type (broad/deep), familiarity (16/8/4/1 experience), and target position (early/late). Accuracy was high across all conditions (95% overall). The results confirmed the expected outcomes that late targets took longer to select and selection was faster as a function of familiarity. The results also showed that selection was significantly faster with the broad version of the menu (10–20 seconds faster, depending on the condition); in fact, "there was clearly no advantage for the deep menus" (p. 317).

Suhm, Freeman, and Getty (2001) studied the effect of long (broad) and short (deep) versions of a touchtone main menu with calls coming into a commercial call center, capturing data from 2834 calls with the broad menu and 2909 calls with the deep version. The broad version had seven options: "For account balance and billing information, press 1. To make a payment by credit card, press 2. If you're having trouble placing or receiving calls in your home area, press 3. For information about placing or receiving calls away from your home area, press 4. For a description of optional features and how to use them, press 5. To discontinue your service, press 6. For all other requests, press 7." The deep version had four options: "For account balance, billing information, or to pay by credit card, press 1. If you're having trouble placing or receiving calls, press 2. For a description of optional features and how to use them, press 3. For all other requests, press 4." As expected, callers using the broad version selected the "Other" choice significantly fewer times than those using the deep version. Contrary to the researchers' expectations, the broad version did not diminish callers' willingness to respond, and the rates of timeouts and invalid responses for broad and deep versions were comparable. The reprompt rate was significantly higher for the deep version (5.1% for deep, 1.7% for broad). Suhm et al. concluded, "Presenting more choices in a menu allows designers of touch-tone voice interfaces to avoid multi-layered menus, which are clearly one of the most dreaded characteristics of touch-tone voice interfaces" (p. 132).

Hura (2010) described the successful use of a 34-option main menu. The application was for internal routing of employees' human-resources calls in a large U.S. corporation. There were two user groups: field users who called several times per day, and home users who called once or twice a year. Previous attempts to group the options into a hierarchical menu structure had been unsuccessful. Hura's solution was to group the main menu into categories that corresponded to a hierarchy, but to eliminate the need to navigate the hierarchy, producing the following very broad main menu (preceded up front by instructions to say a known keyword, followed by an instruction to barge in at any time, with the entire menu spoken in about a minute):

> You can say. … Company Directory. Computer Assistance. Say Field Support to hear those options. For health benefits you can say Unified Healthcare, Rogers, Dental, Healthcare Reimbursement Accounts,

Express Refills, or Other Health Benefits. You can also say Payroll, Direct Deposit, W2, Base Compensation, or Other Payroll Questions. For financial benefits you can say 401K, Trading ID, Stock Options, or Other Financial Benefits. You can say Human Resources, Peopletools, Employment Verification, Job Posting, Termination, or Other HR Issues. For company policies, you can say Family and Medical Leave Act, Short Term Disability, Flexible Work Options, Vacation, Retirement, or Other Policies. Or, you can say Representative if you know you need a person.

Hura (2010, p. 116) concluded, "What we have observed in practice is that frequent users barge in and never hear the menu. Infrequent users who have a term in mind for what they need also barge in without hearing much of the menu. It's only when the user does not know what word to say that they bother listening to the menu. And in that situation, having a menu is an asset, not a punishment. There you have it: thirty four options, quick and simple, the users like it, the client likes it. And I broke a lot of rules."

Commarford et al. (2008) conducted a study to compare broad and deep versions of a speech user interface to e-mail. We started with an existing broad menu that contained 8–11 options presented to callers after having listened to a TTS rendering of an e-mail message. The 11 options were: [Next], [Previous], Repeat, Delete, Reply, List Recipients, [Reply to All], Forward, Mark Unread, Add Sender, Time and Date. Items in brackets only played when applicable (no "Next" for last item, no "Previous" for first item, no "Reply to All" without multiple recipients). Rather than guessing at the appropriate hierarchical organization for the options, we had 26 participants go through a card sorting exercise, analyzed with cluster analysis to group the items, ensuring that no group contained more than five items.

One of the great challenges when building a hierarchical structure is to come up with category labels for the groups. Again, rather than guessing at appropriate category labels, Commarford et al. (2008) conducted two surveys, with 101 respondents to the first survey and 155 to the second. In Survey 1, respondents suggested a label for each group of options. In Survey 2, respondents read descriptions of the menu options, examined each group of options, then using a multiple-choice format, selected the most appropriate label from a list of the most common suggestions generated in Survey 1. The resulting groups of items (and their category labels for the deep menu structure), in order of presentation, were:

- Next, Repeat, Previous (Listen to Messages)
- Reply, Reply to All, Forward, Delete (Respond)
- Add Sender, List Recipients (Distribution)
- Mark Unread, Time and Date (Message Details)

For the broad menu structure, the options played in the order shown above without reference to category labels (Next, Repeat, Previous, Reply, Reply to

All, Forward, Delete, Add Sender, List Recipients, Mark Unread, Time and Date). To provide time for callers to process options, all menu options were separated by 750 ms pauses—consistent with the steady method shown in Figure 8.2. In the final experiment, participants were assigned to the deep or broad version of the application. All participants completed a standardized test of their working memory capacity (WMC) then completed seven tasks with their assigned version of the application. Completion of all tasks required the use of all 11 options at least once. The minimum number of utterances required to complete tasks with the broad version was 33, compared with 62 for the deep version, due to participants having to speak choices from the higher-level menus. The final analyses of results focused on comparisons of performance and satisfaction between the lowest and highest WMC quartiles of participants. The findings showed that it took participants in the deep condition significantly longer to complete the tasks, and those participants also had significantly lower task completion rates and poorer satisfaction ratings. Participants with higher WMC completed significantly more tasks, but the differences between low and high WMC did not lead to significant differences in task time or satisfaction rating. The most important finding was an interaction between menu structure and WMC for task completion time. Specifically, participants with high and low WMC completed tasks at about the same rate when using the broad version (both groups relatively quickly and about equal), but high WMC participants were significantly faster than low WMC participants when using the deep version (both groups relatively slow, but the low WMC group much slower than the high WMC group, by about 30 seconds on average). The results of the final experiment were consistent with the cognitive model of user flow for single-item selection from a single menu shown in Figure 8.5.

Basically, we (Commarford et al., 2008) found that the mental demands of navigating a spoken menu can be greater than the demands of selecting an auditory option from a list, and this most strongly affected the people who were supposed to benefit from shorter menus—people with lower working memory capacity.

- This doesn't mean that designers should pack more options into a menu than will logically fit, but to as great an extent as possible, designers should prefer broad menu structures to deep menu structures.
- Our analysis of working memory demands during selection from an auditory menu and the results of the experiments free SUI designers from the constraint of limiting the number of options per menu to five or fewer, as long as the pacing of presentation of the options provides sufficient time for callers to process the options as they play.
- If the design will have a touchtone-only fallback mode, then the number of choices should not exceed nine (the number of digits

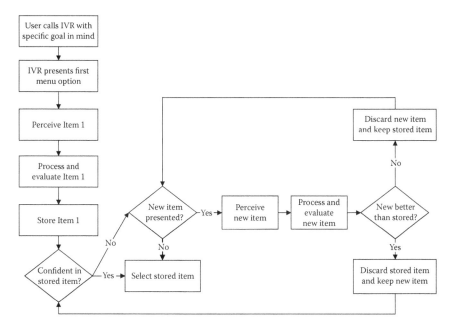

FIGURE 8.5
Cognitive model of user flow for single-item selection from a single menu.

available on a phone keypad—assuming the use of 0 for transfer to agent, # for repeat, and * for go back).

Note that it is the designer's role to determine how best to allow users access to a specified set of options. Although it is an important factor, menu length is not the only factor. It is also important for designers to provide unambiguous menu labels and to place menu items into logical groups to meet user expectations and avoid confusion. If the items fall nicely into groups of four or fewer, it is reasonable to organize them in this manner. The IVR design question under consideration (whether long menus have user performance advantages compared with sets of shorter menus) becomes critical when more than a few items are relevant (potentially useful) at a particular point in the user interface flow. (Commarford et al., 2008, p. 78)

Most recently, Wolters, Georgila, Moore, Logie, MacPherson, and Watson (2009) conducted a "Wizard of Oz" experiment in which 48 participants (26 age 52–84 years; 24 age 18–29) used nine different simulations of speech recognition applications to book appointments. The applications differed in the breadth and depth of their menu structures and the extent to which they required explicit confirmation of caller inputs. Menu structures offered one choice of type of doctor (for example, "Would you like to see the occupational therapist?" selected with "Yes" or "No," four questions deep), two choices

("Would you like to see the occupational therapist or the community nurse?" two questions deep), or four choices ("Would you like to see the occupational therapist, the community nurse, the physiotherapist, or the diabetes nurse?" one question deep). There were three types of confirmation: none, implicit (mentioning the selection as part of the next prompt), and explicit (requiring callers to engage in a separate dialog turn to confirm the input). Wolters et al. (2009) used the same standardized test as Commarford et al. (2008) to assess their participants' WMC (Unsworth & Engle, 2005), which they referred to as WMS (working memory span). There were no significant differences in task completion rates or satisfaction ratings as a function of the independent variables, but when the application presented more options per turn and avoided explicit confirmations (in other words, required the fewest dialog turns), both older and younger participants booked appointments more quickly. Wolters et al. (2009, pp. 283–284) concluded:

> The central challenge in our appointment scheduling task was not to remember all available appointments, but to monitor one's own schedule and detect the option that fits. Since slots were only labeled as free or blocked, users could easily solve the task by scanning, with no need for additional planning. Thus, our results complement Commarford et al.'s (2008) finding that users with a lower WMS benefit from being presented with more options at a time, because at each step in the interaction, they are more likely to be presented with the correct choice. As a result, the overall interaction becomes shorter and less complex.

This literature review covers the studies of which I am aware that have directly addressed the issue of menu breadth and depth, and all have results that favor broad over deep auditory menu structures. There is simply no compelling experimental evidence of improved usability due to shorter menus and deeper structures for either touchtone or speech recognition applications. Two studies published in 2007 jumped through design hoops to try to improve the usability of deep menu structures. Perugini, Anderson, and Moroney (2007) studied "out-of-turn" interaction in menu-based IVR voicemail systems. Basically, they had a relatively deep menu structure (as many as five levels, depending on the task), but activated all options in all levels of the structure at the main menu, letting experienced callers skip levels if they knew what to say. Specifically, the system of Perugini et al. (2007) "retained each sequence through the menus which involves the term in the utterance as a menu item and prunes out all others" (p. 962). Forty participants (20 male, 20 female) used both normal and "out-of-turn" versions of the application in counterbalanced order of presentation to complete a series of voicemail tasks (four experimental tasks with each version). There was no significant difference between the versions with regard to task completion rates, but the other results generally favored the "out-of-turn" version, which had significantly reduced task completion time, higher usability ratings, and a substantial percentage (85%) of participants' preference. This was achieved, however, with a rather complex approach to dynamic menu restructuring. The results suggest that a "talk-ahead" strategy could be advantageous in applications that do not employ natural language understanding (NLU) methods and that have a substantial proportion of high-frequency callers, but do not provide compelling evidence for or against the use of broad over deep menu structures.

Wilkie, McInnes, Jack, and Littlewood (2007) activated a "hidden" overdraft menu option to an existing banking main menu without adding it explicitly as a new option in the menu's list of options. The system informed callers about the new overdraft option in one of three places—in the introduction, immediately after caller identification, and at the completion of the first transaction. For example, the message in the introduction was "We've added a new overdraft facility to this service. To find out more, just say overdraft at the menu of services." The main menu used the "more options" strategy discredited by Virzi and Huitema (1997) to put three options on the initial part of the main menu and four options on the secondary part: "Please select balance, recent transactions or another service"; if "another service," then, "In addition you can select funds transfer, item search, order statement or change TIN. Which service would you like?" The results, based on calls by 114 participants in the study, showed that a significant percentage of participants (37%) did not succeed in completing an overdraft request. Amazingly, in their discussion of future studies, Wilkie et al. (2007, p. 530) stated:

> A perhaps obvious solution to the user's problem (male or female) of obtaining an overdraft would be to simply add an overdraft option to the main menu listing. Using this approach, the purpose of the system-initiated proposal would then be reduced to notifying callers that a new service option has been added, or to advertise particular features pertaining to the new option that may be of interest to the caller. This approach was employed in a follow-up experiment (unpublished results), which employed a similar setup, and resulted in **all participants successfully obtaining an overdraft** [not bold in the original]. However, adding service options to the main menu in this way is not an ideal solution, as it will render listings longer and more cluttered. An alternative method would be to keep the overdraft option hidden, and instead revisit the contents and wording of the system-initiated proposal.

Rather than accepting the results of their unpublished study as evidence that adding the additional menu option would dramatically enhance usability, solving the problem at hand, Wilkie et al. (2007) stated a plan to do future work in increasing the length of their "system-initiated proposal" to "'educate' users that—although the overdraft option is not explicitly listed in the main menu—it can still be accessed by saying the 'overdraft' keyword" (p. 530). It will be interesting to see how those experiments turn out.

Grouping Menu Items

There are two fundamental principles for determining the order in which to present menu options—specific–before–general and frequent–before–infrequent (Balentine & Morgan, 2001). The specificity rule is the more important one because, if presented too early, general options can capture callers' responses before they hear the more specific options. Within that design constraint, frequently selected options should appear earlier than infrequently selected options.

If two options have almost equal specificity and are items that callers might mistake for one another, it might be necessary to place them in a submenu, deepening the menu structure. For example, our usability testing has revealed problems when putting "service locator" and "store locator" in the same main menu, or differentially routing options for "parts" or "accessories." Because the items within the pairs have about equal specificity, some callers select whichever member of the pair is presented first (for example, when the task is to find the nearest store selecting "service locator" before the option for "store locator" plays). In these cases, we've packed the pair into a single option on the main menu ("Parts and Accessories"), with the selection of the target option in a submenu.

Minimizing Error Declarations

It is very common to play error declarations after a caller says something the system couldn't interpret (nomatch event, for example, "I'm sorry, I didn't get that.") or the caller didn't say anything (noinput event, for example, "I'm sorry, I didn't hear that."). There are, however, many potential causes for nomatch and noinput events (Balentine, 2007):

1. Nonhuman background sounds
2. Human speech from background
3. Mix of human speech and nonhuman sounds from background
4. Caller was talking to someone else (primary side conversation)
5. Caller provided meaningless free form speech to application (babble)
6. Caller provided meaningful speech through corrupted channel
7. Caller provided meaningful speech in idiosyncratic fashion
8. Caller confused about current application state
9. Caller provided meaningful speech that should be in grammar but isn't
10. Rejection due to acoustic confusability
11. Caller provided in-grammar speech that the application rejected (false rejection)
12. Caller momentarily distracted from the call

These declarations of error work best for 6 (corrupted channel) and are sensible for 10 and 11 (acoustic confusability, false rejection), but not for the other situations. Shorter declarations ("I'm sorry," "Once again?" "Hmm," "Excuse me?") are better than the longer declarations (Joe, 2007; Tomko & Rosenfeld, 2004), but the best solution is often to skip the declaration altogether—just reprompt. In a study of how operators dealt with speech recognition problems similar to those caused by misrecognition in speech applications, Skantze (2005, p. 338) reported: "Unlike most dialogue systems, the operators did not often signal non-understanding. If they did display non-understanding, this had a negative effect on the user's experience of task success." This strategy (minimizing or omitting error declaration) has published support from EIG usability tests (Balentine, 2007); I've used it in many applications with no indication that it causes usability problems, and it seems to have become a more standard practice among SUI designers. As McKienzie (2008, p. 37) stated, "In the old days of VUI [voice user interface design], we said things like, I'm sorry, I didn't get that. Most of us have moved away from apologizing, but do we need a transition phrase at all? We're finding that getting straight to the reprompt often works better."

Avoiding Touchtone-Style Prompts

Prompts in an application with a touchtone interface typically take the form "For option, do action." With a SUI, however, the option is the action (Balentine, 1999). Thus it is generally better to avoid speech prompts that mimic touchtone-style prompts because these types of prompts are longer than they need to be for most types of menu selections, and can be terribly redundant (Balentine & Morgan, 2001). For example, use:

System: Please select Marketing, <750 ms> Finance, <750 ms> Human Resources, <750 ms> Accounting, <750 ms> or Research.

rather than:

System: For the Marketing department, say Marketing. For Finance, say Finance. For Human Resources, say Human Resources. For Accounting, say Accounting. For Research, say Research.

If the menu items are difficult for a caller to remember (for example, if they are long or contain unusual terms or acronyms), it may be necessary to mimic touchtone prompts, but this would very much be the exception rather than the rule for initial prompts.

Choosing a Complex Alternative

It isn't always possible to have a simple label for a choice. Consider the following prompt:

System: Would you like to hear your account balances at the beginning of every call, or just at the beginning of the first call of the day?

Try to imagine how many different ways a caller might respond to this question. One way to deal with this situation is to change the prompt to a yes or no question that has the form, "You can choose A or B. Would you like A'?" where A and B are complex choices and A' is a short version of A. For example:

System: You can hear your account balances at the beginning of every call, or just at the beginning of the first call of the day. Would you like to hear them in every call?

Crafting Effective Prompts

There are many aspects to consider when deciding how to word an application's prompts and menus. These design choices will have a significant impact on the types of responses callers will provide, and therefore on what grammars or statistical language models must be able to interpret.

Applications that require callers to learn new commands are inherently more difficult to use. Listen to recordings of call center calls, and look at the enterprise's Web site. Make note of the words and phrases that callers typically use to describe common tasks and items, and then use those words and phrases in the application's prompts and grammars.

A major theme that runs through discussions of human communication (Chapter 3), customer service (Chapter 4), and system cost (Chapters 5 and 7) is the importance of efficiency. Thus one of the most important SUI designer skills is the ability to write concise prompts and messages that have an appropriate customer-centered tone (Polkosky, 2005a, 2005b). Some specific techniques for controlling the length of prompts are:

- Avoid lengthy lead-in phrases to a set of options
    - Consider beginning prompts with words like "Select" or "Choose." If using the steady approach to timing the presentation of menu options (see Figure 8.2), avoid the use of "Say" as a lead-in because it does not strongly imply an upcoming set of choices, and we have observed participants in usability studies interpret the pause after the first choice as a turn-taking indication that the first choice is the only option when preceded by "Say."
        - For example, use:
            - System: Please select Checking, <750 ms> Savings, <750 ms> Money Market, <750 ms> or Loans.

- – instead of:
  - – System: Please say Checking, <750 ms> Savings, <750 ms> Money Market, <750 ms> or Loans.
- – or
  - – System: Please choose one of the following: Checking, <750 ms> Savings, <750 ms> Money Market, <750 ms> or Loans.
- When writing conversational prompts strive to keep them concise. For example:
  - – System: Do you want to work with your checking, savings, or money market account?
- Try phrasing a prompt as a question rather than a statement. This often makes the prompt more concise and more conversational.
  - For example, use:
    - – System: Transfer how much?
  - instead of
    - – System: Please say the amount you would like to transfer.
- Do not overuse "please," "thanks," and "sorry." They can be useful and are sometimes necessary, but don't use them automatically or thoughtlessly—use them only when they serve a clear purpose.
- If removing a word will not change the meaning of a prompt or message then consider removing it (while keeping in mind that a certain amount of structural variation in a group of prompts increases the naturalness of the dialog).
- Strive to use clear, unambiguous, and acoustically distinct terms (Suhm, 2008).
- In general, if there is a choice between long and short words that mean essentially the same thing, choose the short word. In most cases the short word will be more common than the long word and callers will hear and process it more quickly. For example, "use" is often a better choice than "utilize" and "help" is often better than "assistance."
- Good prompts do not necessarily use good grammar. In normal conversation many natural phrases do not abide by the rules of grammar, and are often far shorter than their grammatical equivalents. Study transcripts of effective customer–agent interactions to use as models for effective scripting of automated service.
- Use pronouns, contractions, and ellipsis to avoid excessive repetition
  - Ellipsis is the omission of a word or phrase necessary for a complete syntactical construction but not necessary for understanding

- For example, use:
  - System: You have three new messages. The first one's from John Smith. The second is from Jane Doe. The third's from Robert Jones.
- Instead of:
  - System: You have three new messages. The first message is from John Smith. The second message is from Jane Doe. The third message is from Robert Jones.

Managing Digressions

Wilkie, Jack, and Littlewood (2005) reported an experiment in which they studied the effect of contrasting politeness strategies on the effectiveness of system-initiated digressive proposals (interruptions) in human–computer telephone dialogs. "The purpose of making deliberate digressive interruptions in the current experiment was to explore if politeness strategies for human-human interaction … could be employed to mitigate the adverse effects of these dialog intrusions" (p. 49). Their experience illustrates the difficulty of translating research from social psychology and communication theory to specific and effective SUI design. To develop different politeness strategies, they drew upon Brown and Levinson's (1987) theories of politeness, in particular, the concepts of negative and positive face. Negative face refers to an individual's desire to be free to act as one desires. Positive face refers to the desire for approval and appreciation. In conversation, according to Brown and Levinson, speakers can use politeness expressions to avoid threatening the addressee's negative and positive face, and the extent to which a speaker feels the need to be polite differs as a function of the power of the addressee over the speaker and the social distance between speaker and addressee (also see Chapter 3, "Social Considerations in Conversation," for additional discussion of the concept of "face").

Table 8.7 shows the three versions of Wilkie et al.'s (2005, p. 54–55) system-initiated digressive proposals. The positive version focused on being optimistic ("I know you won't mind"), informal ("cutting in"), intensifying interest with the addressee ("special information for you"), exaggerating approval with the addressee ("make your growing savings grow even more"), showing concern for the interests of the addressee ("with your interests in mind, I suggest"), offering and promising ("an On-line Saver account that will give you better interest"), and giving or asking for reasons ("why not set one up today?"). The negative version focused on apologizing ("I'm very sorry to interrupt"), stating the face-threatening act as a general rule ("it is the bank's policy to notify"), impersonalizing speaker and addressee ("notify customers how to"), being indirect ("we wish to inform you"), giving deference ("as a valued customer"), being pessimistic ("you may therefore want to consider"), and going on record as not indebting the addressee ("we would

TABLE 8.7

Digressive Proposals from Wilkie et al. (2005)

Politeness Strategy	Proposal
Positive face-redress (30 seconds duration)	I know you won't mind me cutting in with some special information for you about how to make your growing savings grow even more. We all want the best return possible from our savings. With your interests in mind, I suggest you open an "On-line Saver account" that will give you better interest than the accounts you've got just now. You can transfer money to and from an On-line Saver account through telephone or Internet banking. Why not set one up today? Do you want me to do that for you now?
Negative face-redress (31 seconds duration)	I'm very sorry to interrupt, but it is the bank's policy to notify customers about how to improve their savings returns. We wish to inform you, as a valued customer, that an "On-line Saver account" offers better interest than the accounts you hold at present. You may therefore want to consider opening an account of this type. Transfers to and from On-line Saver accounts are made through telephone or Internet banking. We would be happy to set up an On-line Saver account for you today. Would you like us to do that now?
Bald proposal (18 seconds duration)	I'm interrupting to inform you about how to improve your savings returns. The "On-line Saver account" offers better interest than the accounts you have at present. You can transfer money to and from an On-line Saver account through telephone or Internet banking. Do you want to set up an On-line saver account now?

be happy to"). The bald version focused on avoiding politeness markers to achieve a more direct and concise message ("I'm interrupting you to inform you about") while still maintaining a reasonably professional tone.

The experimental design of Wilkie et al. (2005) allowed a number of interesting analyses. Before undergoing any of the experimental conditions, all 111 participants completed two tasks with the application to establish a usability rating baseline. Following the baseline tasks, 25 participants experienced a third "no proposal" task, similar to the baseline tasks. For their third task, the remaining participants experienced one of the versions of the digressive proposal (29 the positive version, 28 the negative version, and 29 the bald version). The usability ratings remained steady for the "no proposal" group, but fell significantly for all three versions of the digressive proposal, with no significant difference as a function of politeness strategy. Specifically, confronted with the task-interrupting digression, participants rated the task as significantly more frustrating, less enjoyable, less efficient, more in need of improvement, more stressful, more confusing, and more complicated.

As participants in Wilkie et al. (2005) continued in the experiment, those outside of the "no proposal" group, who never experienced any version of the digressive proposals, experienced all three versions in counterbalanced order. After having completed tasks with all three versions, participant ratings indicated that each version achieved its social objectives. The negative version was rated as the most polite, although all three versions received fairly high ratings on that scale (with negative significantly higher than positive, but not significantly higher than bald). The bald version received the best ratings for containing the greatest amount of relevant information and being the shortest and least long-winded. The negative version was rated as more formal, more polite, and more apologetic. Participants rated the positive version as the most manipulative of the three strategies, and rated it as more patronizing and intrusive than the bald version. None of the versions received favorable ratings of efficiency, but when asked which version they preferred, most participants (54%) selected the bald version because "it was shorter and more to the point than the other two" (p. 65). The percentages preferring the positive and negative versions were 11% and 29%, respectively.

Based on the results of this experiment, a reasonable design goal is to avoid these types of messages, especially when they are optional from the caller's point of view. If an enterprise wants to provide this type of information to callers, it is possible to avoid interrupting callers altogether by using a call-center strategy known as a "soft question" (Caras, 2008). The agent waits until he or she has satisfied the caller's immediate need; then, before ending the call, asks a "soft" question, as appropriate, either for purposes of up-selling, cross-selling, or providing useful information.

Sometimes it might be necessary to ask callers to complete a required task before starting the task that they selected. For example, there may be a need for additional security information or to update an address before continuing with the requested task. When crafting these types of messages, focus on caller needs rather than enterprise needs, and try to provide a concise value proposition to the caller. For example:

> System: To protect your account, please answer the following security questions before we continue.

instead of:

> System: To continue, I need you to answer the following security questions.

Stepping Stones and Safety Nets

If a caller doesn't respond successfully to a concise initial prompt, try a stepping stones and safety nets strategy for error recovery. For example, consider the following menu:

> System: Which best describes why you're calling? Select New Claim, Life Policy, or Other Insurance Products.

If callers ask for help (help event), don't respond to the prompt (noinput event) or say something the application can't understand (nomatch event), then follow with help messages such as:

System: Please let us know why you're calling. To report the death or disability of an insured, say New Claim (or press 1). For service or questions on an existing life insurance policy, say Life Policy (or press 2). For all other insurance products, including annuities and mutual funds, say Other Insurance Products (or press 3).

Caller: <help, noinput, or nomatch event>

System: At any time you can say Repeat, Help, Go Back, Start Over, Transfer to Representative, or Goodbye. <2 second pause> To continue, please say New Claim (or press 1), Life Policy (2), or Other Insurance Products (3).

This style lets callers who are successfully engaging with the application move quickly through the task (running across the stepping stones). If a caller isn't successful with the concise initial prompting, the safety net pops open, with:

- Less concise, touchtone-style reprompting
- Presentation of global commands
- Final concise reprompt (with possible switch to touchtone only at this point)

By analogy, this style tries to keep callers on stepping stones because it isn't possible to move quickly in a safety net. If the final attempt is unsuccessful, transfer to an agent if the call center is open; otherwise play the call center hours, provide information about alternative channels of communication (for example, the Internet), and ideally, offer a call-back option.

I've adopted this general strategy based on the available research in SUI design and have found it to work reasonably well. I know of no comparative research, however, on the myriad other possible arrangements of these elements. For example, there might be applications for which it would be better to present the global commands before the less concise reprompts, or to have a very rapid reprompt followed by global commands followed by the touchtone-style reprompting. This is an area in which SUI designers should continue trying different approaches and publishing the results of their experiments. The general strategy of stepping stones and safety nets seems to be solid.

Tips for Voice Spelling

In general, it is wise to avoid dialog steps that require voice spelling. The letters of the alphabet are notoriously difficult for computers (and humans)

TABLE 8.8

Likely Recognition Errors when Spelling

Letter Returned	Most Likely Substituted For	Letter Returned	Most Likely Substituted For
A	I	N	X
B	D	O	L r u
C	V t z	P	d e
D	E b v	Q	U p t
E	V t z	R	I f y
F	S	S	F h j
G	P t v	T	D e g p v
H		U	
I		V	Z
J	A	W	
K	A	X	
L	M	Y	
M	N	Z	V

to recognize (Rolandi, 2006). If a caller must perform voice spelling, then it is possible to take advantage of the fact that some recognition errors are more likely than others. Table 8.8 shows the results of an experiment conducted to investigate patterns of recognition errors for the letters of the English alphabet (Wright, Hartley, & Lewis, 2002). In the table, uppercase letters indicate substitution probabilities that exceeded 10% (relatively high-probability substitutions). Lowercase letters indicate substitutions that occurred during the study, but had substitution probabilities less than 10%. Note that the specific results in the table do not necessarily generalize to other recognizers, but the patterns should be similar, at least roughly, with other recognizers.

For example, if a caller rejects a returned K in a voice-spelling application, then the letter most likely to have actually been spoken is A. Fourteen of the letters in the table have only one substitution for which the probability of substitution exceeded 10%. Eleven of the letters don't have any substitutes for which the probability of substitution exceeded 10%, and six of those didn't have any substitutions at all. This means that whenever the system returns these letters (H, I, U, W, X, and Y), there is a reasonably high likelihood that the speaker actually said that letter. Only two letters (T and G) had two substitutes for which the probability of substitution exceeded 10%. Note that the substitution probabilities are not usually symmetrical (F/S and V/Z are exceptions). If the system returned a D but the caller indicated that it was not correct, the letter the caller said was most likely an E. If the system returned an E but the caller indicated that it was not correct, the letter the caller said was most likely a V.

There are several ways to use this information (or similar information from n-best lists) to improve the caller experience when voice spelling.

If the information that the caller is spelling is unbounded (such as a caller's last name), the system can offer alternatives in the order of their likelihood. For example:

System: Please spell your last name.

Caller: L e w i s

System: L e w i f?

Caller: No.

System: Was the first letter L as in Lucy?

Caller: Yes.

System: Then E as in elephant?

Caller: Yes.

System: Then W as in walrus?

Caller: Yes.

System: Then I as in Iris?

Caller: Yes.

System: Then F as in Frank?

Caller: No.

System: Was it S as in Sally?

Caller: Yes.

System: L e w i s?

Caller: Yes.

If the information that the caller is spelling is bounded (such as a part number, or a list of callers' last names in a database), then it is often possible to detect and correct voice spelling recognition errors without involving the caller. For example, suppose the caller is ordering a part by its code:

System: Part code?

Caller: S X H U F.

System: <Returns F X H U F>, <Checks part number database>, <Doesn't find this number>, <Checks the following possibilities: <F X H U S, S X H U S, S X H U F>, <Only third one is in database>

System: S as in Sam, X, H, U, F as in Frank?

Caller: Yes.

In the example above, the system created the different possible part numbers by using the information from the table of substitutions. Because there were no likely substitutes for X, H, or U, the system left them alone, and systematically changed F to S and vice versa.

It is common for voice spelling modules to accept words from the NATO (or similar) Phonetic Alphabetic—alpha, bravo, Charlie, delta, echo, foxtrot, golf, hotel, India, Juliet, kilo, Lima, mike, November, Oscar, papa, Quebec, Romeo, sierra, tango, uniform, victor, whiskey, x-ray, Yankee, Zulu (Rolandi, 2006). It is, however, important to conduct appropriate grammar testing to ensure that the acoustically distinct members of the phonetic alphabet are also phonetically distinct with anything else active in the grammar. For example, Lewis and Commarford (2003) reported a potential acoustic confusion between "Juliet" and "period" when voice spelling short sentences. For a list of alternative words to consider when creating American English voice spelling grammars; see Lewis (2005).

Davidson, McInnes, and Jack (2004) studied the usability of various dialog design strategies for automated capture of surnames, some of which included spelling the name. Chapter 7 ("Using TTS Prompts") covered some of the usability problems associated with automatic pronunciation of names in speech system output. The problems with system recognition of spoken names are similar. "The first is the large set of names involved in many applications, ranging from a few thousand names to over a million in some cases. The second is the lack of standardized pronunciations for many names; each can have multiple valid pronunciations, which further increases the difficulty of the recognition task" (p. 56).

Davidson et al. (2004) had 95 participants perform tasks in which they were to book themselves and a "friend" on a flight. Including the "friend" allowed the experimenters to exercise some control over the names spoken to the application, and doubled the number of surname recognition attempts. The grammar included 11,926 British surnames (transcribed or inspected by a trained phonetician), modified to ensure that all names spoken in the experiment were in grammar. The participants used three different approaches to providing the surnames: Speak Only ("Please say your surname."), One Stage Speak and Spell ("Please say then spell your surname."), and Two Stage Speak and Spell ("Please say your surname," then, after caller speaks surname, "How do you spell that?"), presumably with all participants using all three methods in a counterbalanced or randomized order. The grammar allowed the use of "double" (as in "H A double-L"), and participants in the One Stage condition were allowed to link the spoken and spelled versions of the surname with "spelt" or "that's" (as in "Jones that's J O N E S").

The dependent measurements in Davidson et al. (2004) were attitude ratings, preference rankings, and successful task completion rates for the flight booking task overall. For all measures, a higher score was better. For Speak Only, the mean attitude rating was 4.57, the task completion rate was 51.6%, and the percentage most preferring the method was 13.7% (95% adjusted–Wald binomial confidence interval ranging from 8.04–22.2%). For One Pass Speak and Spell, the comparable results were 5.18, 80.0%, and 46.3% (95% confidence interval from 36.6–56.3%), and for Two Pass were 5.17, 77.9%, and 37.9% (95% confidence interval from 28.8–48.0%). For these measurements,

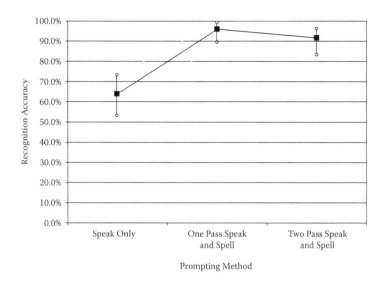

FIGURE 8.6
Recognition accuracies for Speak Only versus Speak and Spell surname entry.

the Speak and Spell methods were consistently superior to the Speak Only method. Although the One Pass method showed slightly better results than the Two Pass method, their outcomes were not significantly different. Many of the task failures for the Speak Only method were due to its substantially lower recognition accuracy. Figure 8.6 shows the recognition accuracies for the 83 participants who spoke their surnames using all three methods, with 95% adjusted–Wald binomial confidence intervals around those means. As the figure shows, the Speak and Spell methods were both significantly better than the Speak Only method, and were not themselves significantly differ-ent (as indicated by the overlapping confidence intervals). The results of this experiment provide strong support for the use of a Speak and Spell strategy when asking callers to provide their name, with the data generally favor-ing the One Pass method. Note, however, that Damper and Gladstone (2007) reported that callers in their usability evaluation had difficulty speaking then spelling their first and last names in the same utterance ("John J O H N Smith S M I T H), so it appears to be necessary to capture first and last names in separate steps if using a Speak and Spell strategy.

Confirming User Input

When a speech application receives input from a caller, it can respond in a variety of ways (McInnes, Nairn, Attwater, Edgington, & Jack, 1999; McTear, O'Neill, Hanna, & Liu, 2005), for example, by:

- Accepting the input and moving on
- Rejecting the input (the nomatch event used in the stepping-stones-and-safety-nets approach to prompting and contextual help)
- Confirming the input, either implicitly or explicitly

Note that there is considerable variability in the terms used in research and practice to describe different confirmation methods. Take care when reading other sources to clearly understand their definitions of terms such as implicit, explicit, blocking move, non-blocking move, feed-forward, and embedded. In this book, "implicit" confirmation refers to confirmation presented in a prompt or message as information related to the input that does not require the caller to accept it explicitly. This contrasts with "explicit" confirmation, in which the caller must respond to a confirmation dialog step before moving on toward task completion.

The most concise dialogs are those that proceed without any confirmation at all, so, all other things being equal, if there is no need to confirm a caller's input, don't (Damper & Gladstone, 2007). When correcting misrecognitions, callers tend to exaggerate (hyperarticulate) the misrecognized portion(s) of the input (Litman, Hirschberg, & Swerts, 2006). Fortunately, this hyperarticulation tends to fade over successive dialog turns without further misrecognition, and the presence of hyperarticulation does not appear to affect recognition accuracy in the affected turns (Stent, Huffman, & Brennan, 2008).

When deciding whether to confirm, it is important to consider the consequence of misrecognition. If the consequences of errors are very low, there may be no need to confirm. Also, sometimes the content of the following prompt or message, without adding any additional information, acts effectively as implicit confirmation. In NL applications that include a dialog manager (see Chapter 2, "Dialog Management"), the dialog manager might make dynamic decisions regarding whether to confirm user input immediately, at a later time in the dialog, or not at all (Krahmer, Swerts, Theune, & Weegels, 2001; Litman, Hirschberg, & Swerts, 2006; McTear et al., 2005; Minker, Pittermann, Pittermann, Strauß, & Bühler, 2007; Torres, Hurtado, García, Sanchis, & Segarra, 2005). For most current IVRs, however, this task falls to the SUI designer.

Implicit Confirmation

Commonly used for high-confidence recognition and situations in which misrecognition does not have serious consequences to the caller (but not quite to the extent that would allow the application to skip confirmation altogether), implicit confirmation feeds key information from the current dialog step forward into the next prompt (Balentine & Morgan, 2001). For example:

System: By Phone or Fax?
Caller: Phone.

System: OK, by phone. Do you want to leave a message or forward the call?

The advantage of implicit confirmation is that it provides feedback that the system correctly understood the response, without the need for cumbersome confirmation of every caller input. The primary disadvantage is that the caller action to disconfirm is not evident. For example, Boyce (2008) reported that only 63% of callers experiencing a simulated error and implicit feedback responded "No" (optionally followed by the correct information), and 15% of callers failed to correct the error at all. Weegels (2000) reported that 8 of 20 participants in a usability study of an application that used implicit confirmation had difficulties responding to implicit confirmations that included errors. Boyce and Viets (2010) have suggested including a quick instruction on what to do as part of implicit confirmation, an approach that seems especially suitable for caller responses that do not have extremely high confidence, for example:

System: Thank you for calling ABC Deliveries. What would you like to do?

Caller: Schedule a pickup.

System: OK, schedule a pickup. If that's not right, say Go Back. Are you shipping air packages that are ready today?

Explicit Confirmation

If the system is about to perform an action it cannot undo, or if there are business rules that require confirmation before execution, then it is reasonable to use explicit confirmation. When confirming, avoid making claims about exactly what the caller said (avoid, for example, "Did you say you wanted to transfer $500 from checking to savings?") and focus instead on task goals (Balentine, 1999). For example:

System: Transfer from?

Caller: My checking account.

System: To?

Caller: My savings.

System: How much?

Caller: $500.

System: OK, you want to transfer $500 from checking to savings, right?

In the example above, the application delayed confirmation until it had all of the information needed to perform the transaction. Explicit confirmation strategies fall into two general categories: immediate confirmation and delayed (or batch) confirmation (Frankish & Noyes, 1990; Kotan & Lewis, 2006).

Immediate Explicit Confirmation: Separate Dialog Turns

The simplest explicit confirmation strategy is to provide a separate confirmation dialog turn for each item. The advantage of this approach is that there is no ambiguity and callers typically experience a very high success rate. The disadvantage is that the additional dialog turns make the call longer and can lead to an interaction that feels sluggish (Boyce, 2008; Frankish & Noyes, 1990). The more items there are the more sluggish the interaction feels. For example:

> System: Which fund?
>
> Caller: Fidelity Magellan.
>
> System: Was that Fidelity Magellan?
>
> Caller: Yes.
>
> System: Buy how many shares?
>
> Caller: 100.
>
> System: Was that 200?
>
> Caller: No.
>
> System: Buy how many shares?
>
> Caller: 100.
>
> System: Was that 100?
>
> Caller: Yes.

Skip Lists

When a caller rejects an item during confirmation, it's usually a good practice to put that response in a skip list. If that item is the most confidently recognized element returned from the grammar after the reprompt, use the skip list to skip over that item and offer the next most confidently recognized element instead. Ainsworth and Pratt (1992, 1993) found that using a skip list, what they called a "repetition with elimination" strategy, required fewer correction steps than their "elimination without repetition" strategy of offering the next most likely recognition candidate when making corrections.

Delayed (Batch) Explicit Confirmation: Basic

The basic delayed confirmation strategy is to play all collected data in a single step, correct any incorrect items, and then repeat the confirmation. The advantage of this approach is that if there are no errors, all confirmation is done in a single step. The disadvantage is that if there are more than two errors, the correction process is very lengthy. The following example illustrates the resulting interaction if there are no errors:

System: Which fund?
Caller: Fidelity Magellan.
System: Buy how many shares?
Caller: 100.
System: Was that 100 shares of Fidelity Magellan?
Caller: Yes.

However, if there are two errors, the resulting dialog is:

System: Which fund?
Caller: Fidelity Magellan.
System: Buy how many shares?
Caller: 100.
System: Was that 200 shares of Fidelity No Load?
Caller: No.
System: Which item do you want to change?
Caller: The number of shares.
System: To how many shares?
Caller: 100.
System: Was that 100 shares of Fidelity No Load?
Caller: No.
System: Which item do you want to change?
Caller: The fund.
System: Change to which fund?
Caller: Fidelity Magellan.
System: Was that 100 shares of Fidelity Magellan?
Caller: Yes.

The more items there are in the confirmation message, the more pronounced this problem can become.

Delayed (Batch) Explicit Confirmation: Improved

To achieve the advantages of delayed confirmation while eliminating the disadvantages of the basic strategy, there are several alternative delayed confirmation methods. The most important aspect of the improved methods is the elimination of the repeated playing of the full confirmation message. If there are no errors, the interaction is the same as that described for the basic delayed confirmation strategy. If there are two errors, the interaction could be:

System: Which fund?
Caller: Fidelity Magellan.
System: Buy how many shares?
Caller: 100.
System: Was that 200 shares of Fidelity No Load?
Caller: No.
System: What do you want to change first?
Caller: The number of shares.
System: To how many shares?
Caller: 100.
System: Was that 100 shares?
Caller: Yes.
System: Change anything else?
Caller: Yes, the fund.
System: Change to which fund?
Caller: Fidelity Magellan.
System: Was that Fidelity Magellan?
Caller: Yes.

If full confirmation (all data) is a customer requirement, do not engage in the immediate confirmation steps. Instead the interaction would be:

System: Which fund?
Caller: Fidelity Magellan.
System: Buy how many shares?
Caller: 100.
System: Was that 200 shares of Fidelity No Load?
Caller: No.
System: What do you want to change first?
Caller: The number of shares.
System: To how many shares?
Caller: 100.

System: Change anything else?

Caller: Yes, the fund.

System: Change to which fund?

Caller: Fidelity Magellan.

System: Was that 100 shares of Fidelity Magellan?

Caller: Yes.

Comparison of Confirmation Strategies

Kotan and Lewis (2006) conducted two experiments investigating these confirmation strategies. The first was an exploratory, small-sample study comparing immediate and basic delayed confirmation. Three participants completed four tasks with a prototype bill-paying application. The tasks differed in confirmation style (immediate or delayed) and number of errors (0 or 2). When there were no corrections required, delayed confirmation was significantly faster than immediate ($t(2) = 16.0$, $p = .004$), but when participants had to make two corrections, immediate confirmation led to faster task completion times ($t(2) = 9.04$, $p = .01$). The results demonstrated a design flaw in the simple delayed confirmation strategy, which required callers to review a fairly large amount of correct input multiple times when making two corrections.

In their second experiment, Kotan and Lewis (2006) compared two different methods for avoiding the design flaw of simple delayed confirmation: Serial Collection/Correction (the method described in the previous section, "Delayed (Batch) Explicit Confirmation: Improved") and Batch Collection/Correction. The Serial method required callers to completely make the first correction (identifying the error and responding to a reprompt for that item of information) before selecting the second item to change. With the Batch method, callers selected all items to change before beginning any correction steps (after selecting the first item to change, callers heard "We'll make that change in a moment. Change anything else?"). Cognitive analysis of the two approaches revealed potential usability issues due to different demands on human working memory when there were two or more items that needed to change. Callers using the Serial method might, during the correction of the first item, either forget the need to change the second item, or forget which other item needed to change. On the other hand, the failure of the Batch method to conform to the stereotype of immediately making corrections after identifying an item to change could disrupt task performance.

Eight participants completed bill-paying tasks in counterbalanced order, one using the Serial method and the other using the Batch method. Six participants preferred the Serial method (observed percentage of 75%, 90% adjusted-Wald binomial confidence interval ranging from 40.0 to 93.7%). More

TABLE 8.9

Recommendations for Design of Delayed Confirmation

Recommendation	Rationale
1. Use the Serial Collection/Correction method for delayed confirmation of multiple items of information in speech recognition IVR applications.	There were fewer task failures when callers experienced the Serial Collection/Correction method. Although not statistically significant, a majority of participants expressed a preference for the Serial method.
2. Provide an option to replay the information if a user has remembered the need to change an additional item but does not remember which item to change.	During post-task interviews, some participants indicated that it would be useful to be able to request a review of the entered data during the correction process. For example, it would be possible to include this option as the final option when listing the items available for change in the confirmation procedure.
3. If an application requires a final review of all data before accepting the data for processing, then do not engage in immediate confirmation during changes. If there is no need for a final comprehensive review of entered data, then use immediate confirmation of changes.	Participants using the new confirmation strategies engaged in immediate confirmation of items after making a change, and never had to review the entire set of entered data. This might be appropriate in some applications, but in other applications (especially financial), business rules might require callers to go through a final review before accepting the data for processing.

important were the successful task completion results. All participants completed the Serial version of the task correctly; three of the eight participants failed when using the Batch version. All three task failures occurred when participants were trying to correct misrecognitions. A statistical test for differences in non-independent tables (Steele & Torrie, 1980) indicated marginal statistical significance (p < .08). Although this is relatively weak empirical evidence, making this is an excellent topic for future research, the combination of the empirical evidence and rational analysis supports the use of the Serial Collection/Correction method for delayed confirmation. Table 8.9 provides the recommendations and rationale from Kotan and Lewis (2006).

Refining Confirmation and Error Correction with Confidence Levels and N-Best Lists

In many cases there is no need to complicate dialog processes with confidence level and n-best information. Indeed, inappropriate use of this information can lead to poor SUI design, for example, assuming that a confidence rating of 0 absolutely means the utterance is out of grammar when this is not necessarily the case (Rolandi, 2003). Sometimes, however, it might be of value to use information about confidence levels and n-best lists to refine confirmation and error correction strategies (Balentine & Morgan, 2001; Cohen, Giangola, & Balogh, 2004).

Confidence levels are values produced by speech recognition engines in which utterances with close matches to recognized words get higher scores. N-best lists are the top n matches produced by a speech recognizer and ranked by confidence level. The match with the highest confidence level is the one that the engine returns to the application.

If using an immediate confirmation strategy with feed forward, don't put the feedback in the following prompt unless the recognition confidence for the item feeding forward is high (Balentine & Morgan, 2001). For example, if confidence is high:

System: Which stock?

Caller: Texaco (recognized with high confidence).

System: How many shares of Texaco?

Caller: 500.

System: Sell 500 shares of Texaco?

Caller: Yes.

But if the recognition confidence is low:

System: Which stock?

Caller: Texaco (recognized as PepsiCo but with low confidence).

System: How many shares?

Caller: 500.

System: One more time, please. What was that stock?

Caller: Texaco (recognized this time with high confidence).

System: Sell 500 shares of Texaco?

Caller: Yes.

Another refinement is to use n-best lists for disambiguation when the top candidates in the n-best list have close confidence scores or are actual homonyms, such as Sysco and Cisco. If it is possible to apply backend logic to disambiguate the candidates, then do so because this saves the caller from needing to disambiguate, reducing the number of dialog steps needed for task completion, thus increasing efficiency. If this isn't possible and it is unlikely that the caller's utterance was out of grammar, provide a disambiguation dialog turn that offers the next top candidate. For example:

System: Do you want to buy Texaco?

Caller: No.

System: Buy PepsiCo?

Caller: Yes.

Summary

To bring a speech IVR application from concept to reality requires a SUI designer to work out all of the low-level, detailed design decisions. Because 100% of an IVR's calls flow through the introduction, it is important to get the introduction right. Introductions should be simultaneously efficient and appropriate to the enterprise's branding. Unless there is a compelling reason to do otherwise (for example, language selection), introductions should start with a welcome message, end with an initial prompt, and include nothing else. In particular, designers should avoid the temptation to load an introduction with Web deflection messages, sales pitches, prompts for touchtone versus speech, or common but ineffective statements such as "Your call is important to us" and "Please listen carefully as our options have changed."

An aspect of detailed design that tends to be underspecified is the timing of the presentation of messages, prompts, and menu options. Even when barge-in is available and callers know they can interrupt system speech, many are reluctant to do so, making it important to provide pauses during system speech to encourage callers who know what to say to go ahead and take their turn to speak. Research on turntaking in human–human conversations over the telephone has shown that pauses of about 1000–1300 ms duration are powerful turntaking cues. The default noinput timeout value for VoiceXML is 7 seconds, with research supporting noinput timeouts as short as 3 seconds, but not shorter. To provide time for callers to process the items in speech menus with more than three options, the inter-item pauses should be about 750–1000 ms in duration. The duration of pauses between directive prompts and prompt extensions (such as quick bits of help on how to respond to the prompt, or at task-terminal points, lists of global navigation instructions) should be about 1500 ms. If there is an expectation that a system will need more than 4 seconds to process an input, the system should acknowledge the input and play a complex processing tone. If the processing time will exceed 15 seconds and a reasonably good estimate is available, it is helpful to announce the expected waiting time.

The use of personal pronouns ("I," "me") in speech application prompts can be acceptable, but it is important to keep the focus of IVR applications on users and their tasks. Prompts can be directive or nondirective, and directive prompts can be highly directive or more conversational. Highly directive prompts tend to lead to highly predictable caller responses, but there is a tradeoff between the level of directness and the extent to which prompts are consistent with customer service behavior and speech characteristics. In general, designers of IVR prompts should attempt to strike a balance between conciseness, clarity, and conversational. To enhance the usability of nondirective prompts, designers should provide appropriate example prompts. For most applications, the example prompts should follow the initial nondirective prompt after a pause of about 2500–3000 ms.

Initial prompts should focus on speech input, even if touchtone input is simultaneously active. An exception to this is the prompting for digit strings such as account numbers or PINs, which can start with the concise lead-in, "Say or enter ...", because for this type of data, about 60% of callers prefer to use the keypad when prompted to do so. Designers should avoid "Press or say <x>" mixed-mode prompting because it inherits the weaknesses of touchtone prompting without taking advantage of the strengths of well-designed SUIs.

Even though they are not especially conversational there are times when auditory menus are appropriate in SUI design. Research since 1997 has shown that the well-known guideline to restrict the number of options in auditory menus to four or five is more of a misguideline. Recent cognitive analyses and empirical evaluations of the auditory menu selection task have shown that deeper menu structures (more menus with fewer options per menu) place more demand on the working memory of callers than broader menu structures (fewer menus with more options per menu). Consequently, SUI designers should not artificially constrain the number of options provided in auditory menus. Two fundamental principles guide the grouping of options in a menu—specific–before–general, then frequent–before–infrequent.

When the application rejects a caller's utterance (a nomatch), do not provide error declarations (such as "I'm sorry, I didn't understand that.") unless there is evidence that the message actually benefits usability for that specific dialog step. Recent research and current practice have shown that such messages rarely provide a benefit to the caller who hears them. Shorter declarations ("Once again?"; "Excuse me?") are better than longer ones, but the best design solution is often to skip the error declaration and move right on to the reprompt.

The crafting of effective prompts is a key SUI design skill. Prompts should promote efficiency while simultaneously providing an appropriate customer-centered tone, either through the wording of the prompt, the way it is spoken, or both. To achieve concise prompts, avoid lengthy lead-in phrases, use questions rather than statements, avoid the overuse of politeness markers, choose shorter rather than longer words, be willing to be ungrammatical as long as the prompt is appropriately conversational, and avoid repetition through the use of pronouns, contractions, and ellipsis. Avoid the use of digressions in introductions or during tasks. To as great an extent as possible, let callers get to and finish their tasks as quickly as possible before providing enterprise messages to the caller, such as up-sell/cross-sell opportunities or information about new services. Use a stepping-stones-and-safety-nets approach to keep callers moving efficiently through the call when things are working well, but providing contextually relevant assistance when there is evidence that the caller is experiencing problems (for example, noinput and nomatch events).

When possible, SUI designers should avoid dialog steps that require voice spelling. Voice spelling is challenging, even for humans speaking to one another over the phone. If voice spelling is the only design solution for a particular dialog step, then designers can take advantage of patterns of acoustic errors and backend processing to help them deal with misrecognitions. For the speech entry of names, research has shown that Speak and Spell strategies have much higher accuracy than Speak Only strategies.

Confirmation of spoken input takes time, so designers should not confirm input unless there are significant consequences to misrecognition in later dialog steps. Confirmation can be implicit or explicit, and explicit confirmation can be immediate or delayed. The problem with implicit confirmation is that if there is a recognition error, then it is not clear what action callers should take to correct the error. If recognition confidence is high and misrecognitions do not have dire consequences, however, SUI designers can consider the use of implicit confirmation. When using explicit confirmation, the best strategy appears to be the Serial Collection/Correction method. Although not always required, it can be of value to use information about confidence levels and n-best lists to refine confirmation and error correction strategies.

References

Ahlén, S., Kaiser, L., & Olvera, E. (2004). Are you listening to your Spanish speakers? *Speech Technology, 9*(4), 10–15.

Ainsworth, W. A., & Pratt, S. R. (1992). Feedback strategies for error correction in speech recognition systems. *International Journal of Man-Machine Studies, 36*, 833–842.

Ainsworth, W. A., & Pratt, S. R. (1993). Comparing error correction strategies in speech recognition systems. In C. Baber & J. M. Noyes (Eds.), *Interactive speech technology: Human factors issues in the application of speech input/output to computers* (pp. 131–135). London, UK: Taylor & Francis.

Attwater, D. (2008). *Speech and touch-tone in harmony [PowerPoint Slides]. Paper presented at SpeechTek 2008.* New York, NY: SpeechTek.

Baddeley, A. D., & Hitch, G. (1974). Is working memory still working? *American Psychologist, 56*, 851–864.

Balentine, B. (1999). Re-engineering the speech menu. In D. Gardner-Bonneau (Ed.), *Human factors and voice interactive systems* (pp. 205–235). Boston, MA: Kluwer Academic Publishers.

Balentine, B. (2006). The power of the pause. In W. Meisel (Ed.), *VUI visions: Expert views on effective voice user interface design* (pp. 89–91). Victoria, Canada: TMA Associates.

Balentine, B. (2007). *It's better to be a good machine than a bad person.* Annapolis, MD: ICMI Press.

Balentine, B. (2010). Next-generation IVR avoids first-generation user interface mistakes. In W. Meisel (Ed.), *Speech in the user interface: Lessons from experience* (pp. 71–74). Victoria, Canada: TMA Associates.

Balentine, B., Ayer, C. M., Miller, C. L., & Scott, B. L. (1997). Debouncing the speech button: A sliding capture window device for synchronizing turn-taking. *International Journal of Speech Technology, 2*, 7–19.

Balentine, B., & Morgan, D. P. (2001). *How to build a speech recognition application: A style guide for telephony dialogues* (2nd ed.). San Ramon, CA: EIG Press.

Barkin, E. (2009). But is it natural? *Speech Technology, 14*(2), 21–24.

Beattie, G. W., & Barnard, P. J. (1979). The temporal structure of natural telephone conversations (directory enquiry calls). *Linguistics, 17*, 213–229.

Blanchard, H. E., & Stewart, O. T. (2004). Conversational re-prompting in natural language dialog. In *Proceedings of the Human Factors and Ergonomics Society 48th annual meeting* (pp. 708–711). Santa Monica, CA: Human Factors and Ergonomics Society.

Boyce, S. J. (2008). User interface design for natural language systems: From research to reality. In D. Gardner-Bonneau & H. E. Blanchard (Eds.), *Human factors and voice interactive systems* (2nd ed.) (pp. 43–80). New York, NY: Springer.

Boyce, S., & Viets, M. (2010). When is it my turn to talk?: Building smart, lean menus. In W. Meisel (Ed.), *Speech in the user interface: Lessons from experience* (pp. 108–112). Victoria, Canada: TMA Associates.

Brems, D. J., Rabin, M. D., & Waggett, J. L. (1995). Using natural language conventions in the user interface design of automatic speech recognition systems. *Human Factors, 37*(2), 265–282.

Brown, P., & Levinson, S. C. (1987). *Politeness: Some universals in language use.* Cambridge, UK: Cambridge University Press.

Bush, R., & Guerra, L. (2006). Yes/no questions are simple, right? In W. Meisel (Ed.), *VUI visions: Expert views on effective voice user interface design* (pp. 93–96). Victoria, Canada: TMA Associates.

Byrne, B. (2003). "Conversational" isn't always what you think it is. *Speech Technology, 8*(4), 16–19.

Caras, R. (April 2008). Why selling means better service and how to get it done properly. *Call Center Times*, pp. 1–3. Retrieved from www.callcentertimes.com/Portals/0/docs/2008april.pdf

Clark, H. H. (1996). *Using language.* Cambridge, UK: Cambridge University Press.

Cohen, M. H., Giangola, J. P., & Balogh, J. (2004). *Voice user interface design.* Boston, MA: Addison-Wesley.

Commarford, P. M., & Lewis, J. R. (2005). Optimizing the pause length before presentation of global navigation commands. In *Proceedings of HCI International 2005: Volume 2—The management of information: E-business, the Web, and mobile computing* (pp. 1–7). St. Louis, MO: Mira Digital Publication.

Commarford, P. M., Lewis, J. R., Al-Awar Smither, J. & Gentzler, M. D. (2008). A comparison of broad versus deep auditory menu structures. *Human Factors, 50*(1), 77–89.

Crystal, T. H., & House, A. S. (1990). Articulation rate and the duration of syllables and stress groups in connected speech. *Journal of the Acoustical Society of America, 88*, 101–112.

Damper, R. I., & Gladstone, K. (2007). Experiences of usability evaluation of the IMAGINE speech-based interaction system. *International Journal of Speech Technology, 9*, 41–50.

Davidson, N., McInnes, F., & Jack, M. A. (2004). Usability of dialogue design strategies for automated surname capture. *Speech Communication, 43*, 55–70.

Dulude, L. (2002). Automated telephone answering systems and aging. *Behaviour and Information Technology, 21*(3), 171–184.

Enterprise Integration Group. (2000). *Speech Recognition 1999 R&D Program: User interface design recommendations final report.* San Ramon, CA: Author.

Foley, J., & Springer, S. (2010). When your caller anticipates you—dialog support for more cooperative conversations. In W. Meisel (Ed.), *Speech in the user interface: Lessons from experience* (pp. 121–125). Victoria, Canada: TMA Associates.

Frankish, C., & Noyes, J. (1990). Sources of human error in data entry tasks using speech input. *Human Factors, 32*(6), 697–716.

Fried, J., & Edmondson, R. (2006). How customer perceived latency measures success in voice self-service. *Business Communications Review, 36*(3), 26–32.

Fröhlich, P. (2005). Dealing with system response times in interactive speech applications. In *Proceedings of CHI 2005* (pp. 1379–1382). Portland, OR: ACM.

Gardner-Bonneau, D. J. (1992). Human factors in interactive voice response applications: "Common sense" is an uncommon commodity. *Journal of the American Voice I/O Society, 12,* 1–12.

Gardner-Bonneau, D. (1999). Guidelines for speech-enabled IVR application design. In D. Gardner-Bonneau (Ed.), *Human factors and voice interactive systems* (pp. 147–162). Boston, MA: Kluwer Academic Publishers.

Gould, J. D., Boies, S. J., Levy, S., Richards, J. T., & Schoonard, J. (1987). The 1984 Olympics message system: A test of behavioral principles of system design. *Communications of the ACM, 30,* 758–569.

Harris, R. A. (2005). *Voice interaction design: Crafting the new conversational speech systems.* San Francisco, CA: Morgan Kaufmann.

Heins, R., Franzke, M., Durian, M., & Bayya, A. (1997). Turn-taking as a design principle for barge-in in spoken language systems. *International Journal of Speech Technology, 2,* 155–164.

Huguenard, B. R., Lurch, F. J., Junker, B. W., Patz, R. J., & Kass, R. E. (1997). Working-memory failure in phone-based interaction. *ACM Transactions on Computer-Human Interaction, 4*(2), 67–102.

Hura, S. L. (2006). Give me a hint. In W. Meisel (Ed.), *VUI visions: expert views on effective voice user interface design* (pp. 79–81). Victoria, Canada: TMA Associates.

Hura, S. L. (2010). My big fat main menu: The case for strategically breaking the rules. In W. Meisel (Ed.), *Speech in the user interface: Lessons from experience* (pp. 113–116). Victoria, Canada: TMA Associates.

Joe, R. (2007). The elements of style. *Speech Technology, 12*(8), 20–24.

Johnstone, A., Berry, U., Nguyen, T., & Asper, A. (1994). There was a long pause: Influencing turn-taking behaviour in human-human and human-computer spoken dialogues. *International Journal of Human-Computer Studies, 41,* 383–411.

Kaiser, L. (2006). The balancing act: Using touchtone and speech. In W. Meisel (Ed.), *VUI visions: Expert views on effective voice user interface design* (pp. 125–129). Victoria, Canada: TMA Associates.

Karsenty, L. (2002). Shifting the design philosophy of spoken natural language dialogues: From invisible to transparent systems. *International Journal of Speech Technology, 5,* 147–157.

Klie, L. (2010). When in Rome. *Speech Technology, 15*(3), 20–24.

Knott, B. A., Bushey, R. R., & Martin, J. M. (2004). Natural language prompts for an automated call router: Examples increase the clarity of user responses. In *Proceedings of the Human Factors and Ergonomics Society 48th annual meeting* (pp. 736–739). Santa Monica, CA: Human Factors and Ergonomics Society.

Kotan, C., & Lewis, J. R. (2006). Investigation of confirmation strategies for speech recognition applications. In *Proceedings of the Human Factors and Ergonomics Society 50th annual meeting* (pp. 728–732). Santa Monica, CA: Human Factors and Ergonomics Society.

Kotelly, B. (2006). Six tips for better branding. In W. Meisel (Ed.), *VUI visions: expert views on effective voice user interface design* (pp. 61–64). Victoria, Canada: TMA Associates.

Krahmer, E., Swerts, M., Theune, M., & Weegels, M. (2001). Error detection in spoken human-machine interaction. *International Journal of Speech Technology, 4,* 19–30.

Lewis, J. R. (2005). Frequency distributions for names and unconstrained words associated with the letters of the English alphabet. In *Proceedings of HCI International 2005: Posters* (pp. 1–5). St. Louis, MO: Mira Digital Publication.

Lewis, J. R. (2007). *Advantages and disadvantages of press or say <x> speech user interfaces* (Tech. Rep. BCR-UX-2007-0002. Retrieved from drjim.0catch.com/2007_AdvantagesAndDisadvantagesOfPressOrSaySpeechUserInter.pdf). Boca Raton, FL: IBM Corp.

Lewis, J. R., & Commarford, P. M. (2003). Developing a voice-spelling alphabet for PDAs. *IBM Systems Journal, 42*(4), 624–638.

Litman, D., Hirschberg, J., & Swerts, M. (2006). Characterizing and predicting corrections in spoken dialogue systems. *Computational Linguistics, 32*(3), 417–438.

Margulies, E. (2005). Adventures in turn-taking: Notes on success and failure in turn cue coupling. In *AVIOS 2005 proceedings* (pp. 1–10). San Jose, CA: AVIOS.

Marics, M. A., & Engelbeck, G. (1997). Designing voice menu applications for telephones. In M. Helander, T. K. Landauer, & P. Prabhu (Eds.), *Handbook of human-computer interaction* (2nd ed.) (pp. 1085–1102). Amsterdam, Netherlands: Elsevier.

Massaro, D. (1975). Preperceptual images, processing time, and perceptual units in speech perception. In D. Massaro (Ed.), *Understanding language: An information-processing analysis of speech perception, reading, and psycholinguistics* (pp. 125–150). New York, NY: Academic Press.

McInnes, F. R., Nairn, I. A., Attwater, D. J., Edgington, M. D., & Jack, M. A. (1999). *A comparison of confirmation strategies for fluent telephone dialogues.* Edinburgh, UK: Centre for Communication Interface Research.

McKienzie, J. (2008). Moving beyond "I'm sorry, I didn't get that." *Speech Technology, 13*(3), 36–38.

McTear, M., O'Neill, I., Hanna, P., & Liu, X. (2005). Handling errors and determining confirmation strategies—an object based approach. *Speech Communication, 45,* 249–269.

Miller, G. A. (1956). The magical number seven, plus or minus two: Some limits on our capacity for processing information. *The Psychological Review, 63,* 81–97.

Minker, W., Pitterman, J., Pitterman, A., Strauß, P.-M., & Bühler, D. (2007). Challenges in speech-based human-computer interaction. *International Journal of Speech Technology, 10,* 109–119.

Perugini, S., Anderson, T. J., & Moroney, W. F. (2007). A study of out-of-turn interaction in menu-based, IVR, voicemail systems. In *Proceedings of CHI 2007* (pp. 961–970). San Jose, CA: ACM.

Polkosky, M. D. (2005a). *Toward a social-cognitive psychology of speech technology: Affective responses to speech-based e-service.* Unpublished doctoral dissertation, University of South Florida.

Polkosky, M. D. (2005b). What is speech usability, anyway? *Speech Technology, 10*(9), 22–25.

Roberts, F., Francis, A. L., & Morgan, M. (2006). The interaction of inter-turn silence with prosodic cues in listener perceptions of "trouble" in conversation. *Speech Communication, 48,* 1079–1093.

Rolandi, W. (2003). When you don't know what you don't know. *Speech Technology, 8*(4), 28.

Rolandi, W. (2004). Rolandi's razor. *Speech Technology, 9*(4), 39.

Rolandi, W. (2005). The impotence of being earnest. *Speech Technology, 10*(1), 22.

Rolandi, W. (2006). The alpha bail. *Speech Technology, 11*(1), 56.

Rolandi, W. (2007a). Aligning customer and company goals through VUI. *Speech Technology, 12*(2), 6.

Rolandi, W. (2007b). The pains of main are plainly VUI's bane. *Speech Technology, 12*(1), 6.

Schumacher, R. M., Jr., Hardzinski, M. L., & Schwartz, A. L. (1995). Increasing the usability of interactive voice response systems: Research and guidelines for phone-based interfaces. *Human Factors, 37,* 251–264.

Sheeder, T., & Balogh, J. (2003). Say it like you mean it: Priming for structure in caller responses to a spoken dialog system. *International Journal of Speech Technology, 6,* 103–111.

Skantze, G. (2005). Exploring human error recovery strategies: Implications for spoken dialogue systems. *Speech Communication, 45,* 325–341.

Steele, R. G. D., & Torrie, J. H. (1980). *Principles and procedures of statistics.* New York, NY: McGraw-Hill.

Stent, A. J., Huffman, M. K., & Brennan, S. E. (2008). Adapting speaking after evidence of misrecognition: Local and global hyperarticulation. *Speech Communication, 50,* 163–178.

Suhm, B. (2008). IVR usability engineering using guidelines and analyses of end-to-end calls. In D. Gardner-Bonneau & H. E. Blanchard (Eds.), *Human factors and voice interactive systems* (2nd ed.) (pp. 1–41). New York, NY: Springer.

Suhm, B., Bers, J., McCarthy, D., Freeman, B., Getty, D., Godfrey, K., & Peterson, P. (2002). A comparative study of speech in the call center: Natural language call routing vs. touch-tone menus. In *Proceedings of CHI 2002* (pp. 283–290). Minneapolis, MN: ACM.

Suhm, B., Freeman, B., & Getty, D. (2001). Curing the menu blues in touch-tone voice interfaces. In *Proceedings of CHI 2001* (pp. 131–132). The Hague, Netherlands: ACM.

Tomko, S., & Rosenfeld, R. (2004). Shaping spoken input in user-initiative systems. In *Proceedings of the International Conference on Spoken Language Processing (ICSLP—Interspeech) 2004* (pp. 2825–2828). Jeju Island, South Korea: ICSA.

Torres, F., Hurtado, L. F., García, F., Sanchis, E., & Segarra, E. (2005). Error handling in a stochastic dialog system through confidence measures. *Speech Communication, 45,* 211–229.

Unsworth, N., & Engle, R. W. (2005). Individual differences in working memory capacity and learning: Evidence from the serial reaction time task. *Memory and Cognition, 33,* 213–220.

Virzi, R. A., & Huitema, J. S. (1997). Telephone-based menus: Evidence that broader is better than deeper. In *Proceedings of the Human Factors and Ergonomics Society 41st annual meeting* (pp. 315–319). Santa Monica, CA: Human Factors and Ergonomics Society.

Voice Messaging User Interface Forum. (1990). *Specification document*. Cedar Knolls, NJ: Probe Research.

Walker, M. A., Fromer, J., Di Fabbrizio, G., Mestel, C., & Hindle, D. (1998). What can I say?: Evaluating a spoken language interface to email. In *Proceedings of CHI 1998* (pp. 582–589). Los Angeles, CA: ACM.

Weegels, M. F. (2000). Users' conceptions of voice-operated information services. *International Journal of Speech Technology, 3*, 75–82.

Wilkie, J., Jack, M. A., & Littlewood, P. J. (2005). System-initiated digressive proposals in automated human-computer telephone dialogues: The use of contrasting politeness strategies. *International Journal of Human-Computer Studies, 62*, 41–71.

Wilkie, J., McInnes, F., Jack, M. A., & Littlewood, P. (2007). Hidden menu options in automated human-computer telephone dialogues: Dissonance in the user's mental model. *Behaviour & Information Technology, 26*(6), 517–534.

Williams, J. D., & Witt, S. M. (2004). A comparison of dialog strategies for call routing. *International Journal of Speech Technology, 7*, 9–24.

Wilson, T. P., & Zimmerman, D. H. (1986). The structure of silence between turns in two-party conversation. *Discourse Processes, 9*, 375–390.

Wolters, M., Georgila, K., Moore, J. D., Logie, R. H., MacPherson, S. E., & Watson, M. (2009). Reducing working memory load in spoken dialogue systems. *Interacting with Computers, 21*, 276–287.

Wright, L. E., Hartley, M. W., & Lewis, J. R. (2002). Conditional probabilities for IBM Voice Browser 2.0 alpha and alphanumeric recognition (Tech. Rep. 29.3498. Retrieved from drjim.0catch.com/alpha2-acc.pdf). West Palm Beach, FL: IBM.

Yankelovich, N. (1996). How do users know what to say? *Interactions, 3*(6), 32–43.

Yudkowsky, M. (2008). The creepiness factor. *Speech Technology, 13*(8), 4.

Yuschik, M. (2008). Silence locations and durations in dialog management. In D. Gardner-Bonneau & H. E. Blanchard (Eds.), *Human factors and voice interactive systems* (2nd ed.) (pp. 231–253). New York, NY: Springer.

Zoltan-Ford, E. (1991). How to get people to say and type what computers can understand. *International Journal of Man-Machine Studies, 34*, 527–547.

9

From "Hello World" to "The Planets": Prototyping SUI Designs with VoiceXML

This chapter teaches methods for using VoiceXML 2.0 to create prototypes of speech user interfaces (SUIs), touching on many, but not all, of the design points discussed in Chapters 7 and 8. It does not teach professional VoiceXML coding practices, is for reference purposes only, and carries no warranty whatsoever. It does, however, teach in a step-by-step fashion critical prototyping skills that are useful for evaluating SUI concepts and in conducting early usability studies. If you're interested in trying out the code or writing your own prototypes, at the time of writing this chapter, there are some free Web-based VoiceXML development services (for example, cafe.bevocal.com or studio.tellme.com) that may be worth exploring.

The primary purpose of VoiceXML is to enable the creation of a spoken user interface to data (see Chapter 2, "VoiceXML"). The design of VoiceXML makes it ideal for coding call flows (system prompts and help messages, speech recognition grammars, and directing the call flow based on user speech). Like HTML, VoiceXML is a computer language that uses tags (also called "elements"). VoiceXML has comparatively few tags—fewer than 50 in Version 2.0. Because VoiceXML uses XML as its foundation, it requires a strict pairing of beginning and ending tags (for example, <tag> paired with </tag>, where the "/" before "tag" indicates "end"). In some cases, it might not be necessary to put any text between a beginning and ending tag, in which case you can combine the beginning and ending tags by putting the "/" at the end of the tag before the closing angle bracket (for example, <tag/>). It is common to forget to end a tag, so keep that in mind when debugging a program.

This chapter uses elements from Version 2.0 of VoiceXML to illustrate ways to create programs that are consistent with current leading practices in the design of speech user interfaces as described in the other chapters in this book. For more information on VoiceXML, see the VoiceXML 2.0 specification (available at www.w3.org/TR/voicexml20/). The remainder of this chapter contains sample programs with explanations of how the various pieces work. For electronic copies of the program examples, see Lewis (2004, available at drjim.0catch.com/vxml20proto-ral.pdf).

```
<?xml version="1.0" encoding="iso-8859-1"?>
<!DOCTYPE vxml PUBLIC "-//W3C//DTD VOICEXML 2.0//EN"
  "http://www.w3.org/TR/voicexml20/vxml.dtd">
<vxml    version="2.0"    xmlns="http://www.w3.org/2001/vxml"
  xmlns:xsi="http://www.w3.org/2001/XMLSchema-instance"
  xsi:schemaLocation="http://www.w3.org/2001/vxml

  http://www.w3.org/TR/voicexml20/vxml.xsd">
<meta  name="GENERATOR"  content="IBM  Voice  Toolkit  for
  Websphere Studio"/>

<form>
  <block>
      Hello, world!
  </block>
</form>

</vxml>
```

FIGURE 9.1
Hello World.

Sample 1: Hello World!

Figure 9.1 shows a traditional "Hello World" program, written in VoiceXML. VoiceXML requires the initial lines (those preceding "<form>") and the last line ("</vxml>"), automatically produced by the voice toolkit I used to create these examples (IBM Voice Toolkit for WebSphere Studio Version 4.2). Note the use of <form> and <block> tags, with text inside the block. Later sections will include more information about these tags. If you ran this program, all it would do is to use the VoiceXML browser's default TTS voice to say "Hello, World," then stop.

Some VoiceXML Concepts

A VoiceXML document forms a conversational finite state (dialog) machine in which the user is always in one dialog at a time. Each dialog determines the next transition. If a URI (uniform resource identifier, such as a dialog path pointing to a part of the current VoiceXML file, to another VoiceXML file in the same directory, or to a VoiceXML file on a server) does not refer to a document, the system assumes the current document. If it does not refer to a dialog, the system assumes the first dialog in the document. Execution ends when a dialog does not specify a next dialog (as in the Hello World! Sample), or if the dialog explicitly exits the conversation (using an exit tag—<exit/>).

There are two types of VoiceXML dialogs: menus and forms. In this chapter I only use forms. Menus have some interesting properties, but these properties can make them more complicated to use and I have not found them necessary in the prototypes I've built.

Within a form, the key elements are blocks and fields:

- Use blocks to specify actions that do not require interaction with the user.
- Use fields to specify actions that do require interaction (dialog) with the user.

By default, the barge-in type for VoiceXML 2.0 is "hotword," which means that system speech will not stop unless the application receives valid input from the user. You can improve usability by changing the barge-in type to "speech," which means that system speech stops as soon as the application detects speech energy. Both barge-in types have their strengths and weaknesses (see Chapter 7, "Choosing the Barge-In Style"). The sample programs in this chapter will use the "speech" barge-in type, which is set using a property tag:

```
<property name="bargeintype" value="speech"/>
```

Sample 2: Hello Worlds

This sample (Figure 9.2) introduces interaction with a user. Note that it includes the <property> tag that sets the barge-in method to "speech." When I wrote the program Pluto, now "officially" a Kuiper belt object, was an official planet (en.wikipedia.org/wiki/Pluto).

Let's take a closer look at the first five lines of the form. If you want to be able to specify transitions to different parts of your VoiceXML application, you need to provide labels for the various parts. Forms have IDs—blocks and fields have names. You can use single or double quotes around IDs and names. Within the first five lines of the form, there are labels for the form ("helloworlds"), block ('introduction'), and field ('planet'), and the text for the introduction has been specified ("Thank you for calling Hello Worlds!").

```
<?xml version="1.0" encoding="iso-8859-1"?>
<!DOCTYPE vxml PUBLIC "-//W3C//DTD VOICEXML 2.0//EN"
    "vxml20-1115.dtd">
<vxml version="2.0" xmlns="http://www.w3.org/2001/vxml">
<meta name="GENERATOR" content="Voice Toolkit for Websphere
    Studio"/>

<property name="bargeintype" value="speech"/>

<form id="helloworlds">
    <block name='introduction'>
        Thank you for calling Hello Worlds!
    </block>
    <field name='planet'>
        <prompt>
            Which planet? <break time="3s"/> Select Mercury, Venus,
                Earth, Mars, Jupiter, Saturn, Uranus, Neptune, or Pluto.
        </prompt>

        <grammar version="1.0" mode="voice" root="planet">
            <rule id="planet" scope="public">
                <one-of>
                    <item>Mercury</item>
                    <item>Venus</item>
                    <item>Earth</item>
                    <item>Mars</item>
                    <item>Jupiter</item>
                    <item>Saturn</item>
                    <item>Uranus</item>
                    <item>Neptune</item>
                    <item>Pluto</item>
                </one-of>
            </rule>
        </grammar>
        <filled>
            Hello, <value expr="planet"/>! Goodbye.
        </filled>
    </field>
</form>

</vxml>
```

FIGURE 9.2
Hello Worlds.

Now let's examine the next three lines of the form (the prompt component in the field):

```
<field name='planet'>
    <prompt>
        Which planet? <break time="3s"/> Select Mercury, Venus, Earth,
            Mars, Jupiter, Saturn, Uranus, Neptune, or Pluto.
    </prompt>
```

This example uses a short initial prompt (relatively open-ended question), followed by a 3-second pause (break tag), followed by an explicit listing of choices.

If the user knows the names of the nine planets of the solar system (as most users would), the short initial prompt provides a shortcut for getting to the next step of the program. If a user does not know what to say, there is only a 3-second delay before the explicit presentation of the planet's names.

The inline grammar specifies the words/phrases that the recognizer will accept. One attribute of the grammar tag specifies the root (initial, or beginning) rule for the grammar. This grammar has only one rule, labeled with the ID 'planet.' Inside that rule, the choices are the items that appear between of <one-of> and </one-of> tags.

```
<grammar version="1.0" mode="voice" root="planet">
    <rule id="planet" scope="public">
        <one-of>
            <item>Mercury</item>
            <item>Venus</item>
            <item>Earth</item>
            <item>Mars</item>
            <item>Jupiter</item>
            <item>Saturn</item>
            <item>Uranus</item>
            <item>Neptune</item>
            <item>Pluto</item>
        </one-of>
    </rule>
</grammar>
```

The last part of the form is the "filled" section. This is the part of the form in which you specify what to do once the user has filled a field with a valid response to a prompt.

```
<filled>
    Hello, <value expr="planet"/>! Goodbye.
</filled>
```

When filled, the application plays "Hello," followed by whatever value filled the field with the name "planet," followed by "Goodbye." Note the use of the <value> element to feed back the planet name selected by the user. When you give a field a name, you are implicitly declaring a variable with that name to hold the value provided by the user.

Sample 3. Adding More Complex Features to Hello Worlds

In this section, we'll add the following features to the existing Hello Worlds program (Figure 9.3):

- Use a variable and an <if> statement to determine whether to play the introduction
- Set up two always-active commands using <link>
- Use grammar tags to group planets into classes
- Use <assign> to reassign values to variables
- Give variables document-level scope
- Use <if> and <elseif> to direct the call flow

Playing an Introduction Once

First, let's look at the beginning of the program.

```
<var name="skipintro" expr="'play'"/>
```

This line declares a variable ("skipintro"), used to decide whether or not to play the introduction (see the 'if' in the introduction, located in the first block of the form). Note that the expr attribute gives the variable its initial value ('play'). ("Expr" is an abbreviation of the word "expression.")

In VoiceXML, if the value assigned to a variable is enclosed in single quotes inside a pair of double quotes, then it is a literal value (exactly, or literally, the character string enclosed in the single quotes). For example, in the <var>

```
<?xml version="1.0" encoding="iso-8859-1"?>
<!DOCTYPE vxml PUBLIC "vxml" "">
<vxml version="1.0">

<property name="bargeintype" value="speech"/>

<var name="skipintro" expr="'play'"/>

<link next="#helloworlds">
    <grammar version="1.0" mode="voice" root="returntomain">
        <rule id="returntomain" scope="public">
            <one-of>
                <item>main menu</item>
                <item>start over</item>
            </one-of>
        </rule>
    </grammar>
</link>

<link next="#exit">
    <grammar version="1.0" mode="voice" root="goodbye">
        <rule id="goodbye" scope="public">
            <one-of>
                <item>goodbye</item>
                <item>exit</item>
            </one-of>
        </rule>
    </grammar>
</link>

<form id="helloworlds">
    <block name='introduction'>
        <if cond="skipintro == 'skip'">
            <goto nextitem="planet"/>
        </if>
        Thank you for calling Hello Worlds!
    </block>
```

continued

FIGURE 9.3
Hello Worlds Version 3.

```
<field name='planet'>
    <prompt>
        Which planet? <break time="3s"/> Select Mercury, Venus,
            Earth, Mars, Jupiter, Saturn, Uranus, Neptune, or Pluto.
    </prompt>
    <grammar version="1.0" mode="voice" root="planet">
        <rule id="planet" scope="public">
            <one-of>
                <item>Mercury<tag>$="inner"</tag></item>
                <item>Venus<tag>$="inner"</tag></item>
                <item>Earth</item>
                <item>Mars<tag>$="outer"</tag></item>
                <item>Jupiter<tag>$="outer"</tag></item>
                <item>Saturn<tag>$="outer"</tag></item>
                <item>Uranus<tag>$="outer"</tag></item>
                <item>Neptune<tag>$="outer"</tag></item>
                <item>Pluto<tag>$="outer"</tag></item>
            </one-of>
        </rule>
    </grammar>
    <filled>
        <assign name="document.planettype" expr="planet"/>
        <assign name="document.planet" expr="planet$.utterance"/>
        <assign name="document.skipintro" expr="'skip'"/>
        <if cond="planettype == 'inner'">
            <goto next="#innerworld"/>
            <elseif cond="planettype == 'outer'"/>
            <goto next="#outerworld"/>
            <else if cond="planettype == 'Earth'"/>
            <goto next="#earth"/>
        </if>
    </filled>
</field>
</form>
```

continued

FIGURE 9.3 (continued)
Hello Worlds Version 3.

```
<form id='earth'>
   <block>
       <value expr="planet"/> is home, sweet home.
       <goto next="#helloworlds"/>
   </block>
</form>


<form id='innerworld'>
   <block>
       <value expr="planet"/> is an <value expr="planettype"/>
          planet.
       <goto next="#helloworlds"/>
   </block>
</form>


<form id='outerworld'>
   <block>
       <value expr="planet"/> is an <value expr="planettype"/>
          planet.
       <goto next="#helloworlds"/>
   </block>
</form>


<form id='exit'>
   <block>
       Thanks for calling Hello Worlds. Goodbye!
       <exit/>
   </block>
</form>


</vxml>
```

FIGURE 9.3 (continued)
Hello Worlds Version 3.

element used in this version of the program, the value given to skipintro is the character string 'play.' If you're assigning to a variable the value currently assigned to a different variable, then enclose the variable name in double quotes only. For example, suppose play was a variable that currently had the literal value 'yes,' and you wanted to assign to the variable skipintro the current value of play. In that case, the code for declaring the variable skipintro would have looked like <var name="skipintro" expr="play"/>, and the current literal value of skipintro would be 'yes.' If a value is numeric rather than character (i.e., you plan to use it in numeric operations), then enclose it only in double quotes (for example, "100").

Next, let's go through the beginning of the first form:

```
<form id="helloworlds">
   <block name='introduction'>
      <if cond="skipintro == 'skip'">
         <goto nextitem="planet"/>
      </if>
      Thank you for calling Hello Worlds!
   </block>
<field name='planet'>
```

The <if> statement in the introduction block skips the introduction if the value of skipintro is 'skip,' and transfers the call flow to the item named 'planet' (which is the name of the field). Because the initial value of skipintro is 'play,' the introduction will play the first time the program runs this form. Later, we'll change the value to 'skip' to prevent it from playing every time.

Note that the <if> statement has two key parts—the cond (short for condition) and <goto>. Pay careful attention to the use of quotation marks in the cond attribute. The interpretation of this cond attribute is "If the variable named skipintro has a value of 'skip,' then execute whatever code is between <if> and </if>. Otherwise, ignore the code between <if> and </if> and continue with the code that follows </if>." In this case, the code that is between <if> and </if> is a <goto> that transfers the call flow to the item in the current form (indicated by the use of 'nextitem') that has the name 'planet.' The line of code that follows the </if> is the line that plays the introduction ("Thank you for calling Hello Worlds!"). When the condition for the <if> statement is true, the program skips the introduction.

Defining Always-Active Commands with Links

The link elements define always-active commands for (1) returning to the main menu and (2) stopping the program. If a user says "main menu" or "start over" the call flow immediately goes to the form with the ID 'helloworlds.'

If a user says "goodbye" or "exit," the call flow goes immediately to a form with the ID 'exit.' When the target for transfer defined with the 'next' attribute starts with "#", the target is the ID of a form in the current document. Otherwise VoiceXML interprets the target as the name of a completely different document.

```
<link next="#helloworlds">
    <grammar version="1.0" mode="voice" root="returntomain">
        <rule id="returntomain" scope="public">
            <one-of>
            <item>main menu</item>
            <item>start over</item>
            </one-of>
        </rule>
    </grammar>
</link>

<link next="#exit">
    <grammar version="1.0" mode="voice" root="goodbye">
        <rule id="goodbye" scope="public">
            <one-of>
            <item>goodbye</item>
            <item>exit</item>
            </one-of>
        </rule>
    </grammar>
</link>
```

Using Grammar Tags to Classify Responses

The rest of the form is the same as the previous version of Hello Worlds, up to the grammar:

```
<grammar version="1.0" mode="voice" root="planet">
    <rule id="planet" scope="public">
        <one-of>
            <item>Mercury<tag>$="inner"</tag></item>
            <item>Venus<tag>$="inner"</tag></item>
            <item>Earth</item>
```

```
            <item>Mars<tag>$="inner"</tag></item>
            <item>Jupiter<tag>$="outer"</tag></item>
            <item>Saturn<tag>$="outer"</tag></item>
            <item>Uranus<tag>$="outer"</tag></item>
            <item>Neptune<tag>$="outer"</tag></item>
            <item>Pluto<tag>$="outer"</tag></item>
         </one-of>
      </rule>
   </grammar>
```

The new part of the grammar is that the planets (except for 'Earth') now have tags (the character strings enclosed in structures like <tag>$="string"</tag>) that classify the planet as an inner or outer planet (inside or outside the asteroid belt). When grammar tags are present, they replace the recognized words as the value of the variable implicitly declared by naming the field (in this case, the variable named 'planet'). Grammar tags can be very useful for putting things into classes or for indicating that different words or phrases mean the same thing (which is a type of classification).

Setting Variables to Document-Level Scope

The first three lines in the <filled> section take care of some assignment of values to variables for use in other forms. For this reason, all of the variable names start with 'document.' The default scope of a variable in VoiceXML depends on the place in the program where the variable was declared. If declared in a form, then the variable will exist only while that form is active. Once the call flow leaves that form, the variable ceases to exist. To prevent this from happening, you can put 'document' as the first part of the variable name during assignment, and the variable will have document-level scope instead of the default form-level scope. This means that the variable will continue to exist and be available for use anywhere in the document. Because skipintro was declared outside of a form, it has document-level scope by default (but it doesn't hurt to include 'document' when referring to it).

```
   <filled>
      <assign name="document.skipintro" expr="'skip'"/>
      <assign name="document.planettype" expr="planet"/>
      <assign name="document.planet" expr="planet$.utterance"/>
```

Specifically, the first line changes the value of skipintro to 'skip' and sets skipintro to document-level scope (so from now on, if the call flow transfers back to the first form 'helloworlds,' the program will skip the introduction). The second line assigns to a new variable named planettype (set to document-level scope) the value of the variable named planet (which will be 'inner,' 'outer,' or 'Earth'). The third line assigns to the variable named planet (also set to document-level scope) whatever the grammar has recorded as what the user actually said rather than the value of the associated grammar tag (using the shadow variable "planet$.utterance"). In VoiceXML, 'shadow' variables are variables that the system generates automatically. To use a shadow variable, you must use the correct syntax. For this shadow variable, the required syntax is to type the variable name to which you are referring (in this example, 'planet'), followed by a dollar sign, followed by '.utterance'.

Directing the Call Flow with If Statements in the Filled Section

The <if> statement in the <filled> section is more complex than the first <if> statement because it has multiple conditions and transfers (defined with <elseif>).

```
<if cond="planettype == 'inner'">
   <goto next="#innerworld"/>
   <elseif cond="planettype == 'outer'"/>
   <goto next="#outerworld"/>
   <elseif cond="planettype == 'Earth'"/>
   <goto next="#earth"/>
</if>
</filled>
```

Depending on the value of planettype (which can be 'Earth,' 'inner,' or 'outer'), the call flow goes to forms with the IDs 'earth,' 'innerworld,' or 'outerworld,' respectively.

Using the Document-Level Variables and Unconditional Return to the Main Menu

Next, let's take a look at those three forms:

```
<form id='earth'>
   <block>
      <value expr="planet"/> is home, sweet home.
      <goto next="#helloworlds"/>
   </block>
</form>
```

```
<form id='innerworld'>
   <block>
      <value expr="planet"/> is an <value expr="planettype"/> planet.
      <goto next="#helloworlds"/>
   </block>
</form>

<form id='outerworld'>
   <block>
      <value expr="planet"/> is an <value expr="planettype"/> planet.
      <goto next="#helloworlds"/>
   </block>
</form>
```

All three forms have the same structure, containing a single block that contains a statement followed by a <goto> that returns the call flow to the form named 'helloworlds.' All of the statements use <value expr="planet"/> to play the name of the selected planet. The statements in the innerworld and outerworld forms also use <value expr="planettype"/> to play the type of planet (inner or outer).

Exiting without Confirmation

The last part of the program is the exit form, shown below:

```
<form id='exit'>
   <block>
      Thanks for calling Hello Worlds. Goodbye!
      <exit/>
   </block>
</form>
```

This is a simple form that plays a goodbye sentence ("Thanks for calling Hello Worlds. Goodbye!"), followed by an <exit/> element, which stops the program. Note that this form does not give users an opportunity to stop the exit and return to the application—a practice that I would not normally advocate (and which we take care of in Version 4).

Sample 4. Even More Features

In this section, we'll add the following new features to Hello Worlds:

- Self-revealing help
- Go back (unconditional)
- Go back (conditional)
- Exit with confirmation
- Audio tones
- Simulated data acquisition from a backend server
- Playing the data
- Use of breaks to fine-tune timing

Figure 9.4 (which appears at the end of this section) contains the entire program. Before trying to work your way through the whole thing, let's go over the key new features, starting with self-revealing help.

Creating Self-Revealing Help

Self-revealing help refers to a help strategy in which you reveal more contextually relevant information at a dialog turn each time the user says "Help," says something the system can't understand (called, in VoiceXML, a nomatch event), or doesn't say anything for a defined silence timeout period (usually 7 seconds—called, in VoiceXML, a noinput event). For applications that can automatically transfer to a call center staffed with human agents, it is common to provide two levels of help, followed by a transfer to an agent on the third nomatch or noinput event. For applications that cannot transfer to a human agent, one strategy is to cycle through the two help levels. Note that this is an untested strategy for agentless systems, and early usability testing suggests that it can lead to usability problems when users don't realize that they're cycling between two help levels. It might be necessary to modify this strategy in a real system, possibly by labeling the help levels (which is, again, an untested strategy). Please be aware of this if you plan to build a system using this method, and be sure to conduct usability studies to make sure that it works for your application.

Here is an example of self-revealing help from the main menu of the Hello Worlds application, Version 4:

```
<var name="helpcounter" expr="0"/>
... first lines of helloworlds form, through the end of the prompt ...
```

```
<catch event="help noinput nomatch">
    <assign name="helpcounter" expr="helpcounter+1"/>
    <if cond="helpcounter == 1">
        <prompt>
            <break time="150ms"/>Please say the name of a planet.
            <break time="2s"/> Select Mercury, Venus, Earth, Mars,
                Jupiter, Saturn, Uranus, Neptune, or Pluto.
        </prompt>
    </if>
    <if cond="helpcounter == 2">
        <prompt>
            <break time="150ms"/>At any time you can say Repeat, Help,
                Go Back, Start Over, or Exit. To continue, say Mercury,
                Venus, Earth, Mars, Jupiter, Saturn, Uranus, Neptune, or
                Pluto.
        </prompt>
        <assign name="helpcounter" expr="0"/>
    </if>
</catch>
```

When declared, the variable named helpcounter gets an initial value of 0. The <catch> element is set to respond to help, noinput, and nomatch events. The <assign> statement under <catch> increments helpcounter to play the next level of help. The line at the end of the second help level resets help-counter to 0. The effect is to cycle between the two levels of help. (To proto-type transfer, you could use a third level of help that transfers the call flow to a fake transfer form.) To ensure that the help messages in a form will play in the correct order, it is important to reset the value of helpcounter by includ-ing a line of code at the beginning of the form that assigns it the value 0, as shown below:

```
<form id='gettopic'>
    <block>
        <assign name="helpcounter" expr="0"/>
    </block>
```

Defining an Unconditional Go Back

It can be handy for users to be able to work their way back through a call flow using an always-active 'go back' command. It's pretty easy to do this for simple call flows, but it requires the following types of code:

- The declaration of a variable named goback.
- A link that defines the always-active 'go back' command.
- Code in every form containing a field that defines the place to which the program should transfer the call flow if a user says "go back."
- A form that actually performs the 'go back' function.

Here's the implementation of 'go back' in Hello Worlds Version 4, starting with the declared variable (with an initial value of 'undefined') and the link (which programs the application to go to a form with the ID 'goback' if the user says "go back"):

```
<var name="goback"/>
<link next="#goback">
    <grammar version="1.0" mode="voice" root="return">
        <rule id="return" scope="public">
            <one-of>
            <item>go back</item>
            </one-of>
        </rule>
    </grammar>
</link>
```

Hello Worlds Version 4 prompts users for two pieces of information—the planet (like the previous versions) and a planetary attribute, such as temperature, number of moons, distance from the sun, etc. If a user has provided a planet for the first prompt, then, on hearing the second prompt decides to change the planet, he or she can say "go back," as long as the transfer target for 'go back' has been defined. One way to take care of this is to assign the 'go back' target in a block placed at the beginning of each form. Here's the code that does this at the beginning of the Version 4 form that prompts the user for a planetary attribute (the 'gettopic' form):

```
<form id='gettopic'>
    <block>
        <assign name="document.goback" expr="'helloworlds'"/>
    </block>
```

So, if a user is in the 'gettopic' form and says "go back" the call flow will transfer back to the 'helloworlds' form after the processing defined in the 'goback' form:

```
<form id='goback'>
    <block>
        <goto expr="'#'+document.goback"/>
    </block>
</form>
```

All the 'goback' form does is to use an expr attribute to concatenate a '#' with the current value of the goback variable inside of a <goto> statement, which has the effect of going back to the previous dialog.

Defining a Conditional Go Back

This 'go back' strategy works well as long as the structure of the application is purely hierarchical—in other words, there's only one place to go back to from any other place. If it's possible to get to a place in the user interface from more than one path, then the 'go back' strategy for that place will need to be conditional.

For example, imagine an application in which a user can check on the current balance of a bill by making a selection directly from the main menu, or can first select Billing from the main menu, then Current Balance. The conditional goback in the form that plays the current balance could have the following structure:

```
<form id='currentbalance'>
    <block>
        <if cond="mainChoice == 'current balance'">
            <assign name="document.goback" expr="'mainMenu'"/>
        <elseif cond="mainChoice == 'billing'"/>
            <assign name="document.goback" expr="'billing'"/>
        </if>
    </block>
```

This code specifies that if the user's most recent choice from the main menu (mainChoice) was 'current balance,' then the value of goback is 'mainMenu' (returning the user to the Main Menu). If the user's most recent choice from the main menu was 'billing,' then the value of goback is 'billing' (returning the user to the Billing Menu).

Exiting with Confirmation

If the system activates the exit command, you might want to have the user provide confirmation in case:

- The user actually said something else
- The user didn't realize that it would end the call

If the user rejects the confirmation, then the application should return the user to the form in use when the system detected the exit command. One way to do this is to use a strategy similar to the one described for 'go back.' The components needed to accomplish this are:

- The declaration of a variable named currentform.
- A link that defines the always-active 'exit' command.
- Code in every form containing a field that assigns the form's ID to a variable.
- An exit confirmation form.

Here is the declaration of currentform in Hello Worlds Version 4, followed by the link that establishes the always-active exit command:

```
<var name="currentform"/>
<link next="#confirmexit">
    <grammar version="1.0" mode="voice" root="goodbye">
        <rule id="goodbye" scope="public">
            <one-of>
            <item>goodbye</item>
            <item>exit</item>
            </one-of>
        </rule>
    </grammar>
</link>
```

In this version of the program, users can say "goodbye" or "exit" to jump to the exit confirmation form.

Next is an example of the assignment of a form's ID to a variable:

```
<form id='gettopic'>
    <block>
        <assign name="currentform" expr="'gettopic'"/>
    </block>
```

Finally, here is the exit confirmation form from Hello Worlds Version 4:

```
<form id='confirmexit'>
    <field name="exitChoice" type="boolean">
        <prompt>
            <break time="150ms"/>
            Do you want to end this call?
        </prompt>
        <catch event="help noinput nomatch">
        <assign name="helpcounter" expr="helpcounter+1"/>
        <if cond="helpcounter == 1">
            <break time="150ms"/>Please say Yes, No, or Repeat.
        </if>
        <if cond="helpcounter == 2">
            <break time="150ms"/>At any time you can say Repeat, Help,
                Go Back, Start Over, or Exit. To end the call, say Yes.
            To return to Hello Worlds, say No.
            <assign name="helpcounter" expr="0"/>
        </if>
        </catch>
        <filled>
        <if cond="exitChoice">
            <goto next="#exit"/>
            <else/>
            <audio src="triple.au"/>
            <break time="150ms"/>
            Returning.
            <goto expr="'#'+document.currentform"/>
        </if>
        </filled>
    </field>
</form>
```

Note the use of the Boolean (yes/no) grammar type (specified as an attribute in the <field> tag) and the special form of the comparison for a Boolean value (<if cond="exitChoice">). This has the same meaning as <if cond="exitChoice == 'yes'">. If the caller has said "yes" (or any synonym for "yes" available in the Boolean grammar), the program goes to the form with the ID of 'exit.' Otherwise, the program uses the same technique as that described for 'go back' to return to the form that was current when the exit procedure started (<goto expr="'#'+document.currentform"/>).

Integrating Audio Tones

You can use the <audio> tag to put tones in an application. This example shows the use of a tone named triple.au (three rapid beeps) to provide audio highlighting (both before and after playing the planet name) for a planet landmark:

```
<form id="playlandmark">
    <block>
        <audio src="triple.au"/>
        <break time="150ms"/>
        <value expr="document.planet"/>
        <break time="150ms"/>
        <audio src="triple.au"/>
        <goto next="#gettopic"/>
    </block>
</form>
```

Simulating Data Acquisition from a Backend Server

In a prototype, it might be necessary to simulate the acquisition of data from a database on a backend server. The following lines illustrate an example of this:

```
<form id='lookupdata'>
    <block>
        <if cond="document.planet == 'mercury'">
            <assign name="document.distance" expr="'58 million
                kilometers'"/>
            <assign name="document.position" expr="'the closest planet
                to the sun'"/>
            <assign name="document.orbit" expr="'88 days'"/>
            <assign name="document.temperature" expr="'440 degrees
                Celsius'"/>
```

```
        <assign name="document.size" expr="'next to smallest'"/>
        <assign name="document.atmosphere" expr="'98% helium,
            2% hydrogen'"/>
        <assign name="document.moons" expr="'no moons'"/>
    </if>
    …
    <if cond="document.planet == 'pluto'">
        <assign    name="document.distance"    expr="'5.9    billion
            kilometers'"/>
        <assign name="document.position" expr="'usually the far-
            thest planet from the sun'"/>
        <assign name="document.orbit" expr="'248 and a half years'"/>
        <assign name="document.temperature" expr="'negative 233
            degrees Celsius, only 40 degrees above absolute zero'"/>
        <assign name="document.size" expr="'smallest'"/>
        <assign name="document.atmosphere" expr="'mostly nitro-
            gen, with some carbon monoxide and methane'"/>
        <assign name="document.moons" expr="'one moon'"/>
    </if>
```

Playing the Data

After getting the data, you need to play it, as shown in the following lines of code (note the use of document-level variables):

```
<form id='playit'>
    <block>
        <if cond="document.topic == 'distance'">
            At an average distance of <value
            expr="document.distance"/>,
                <value expr="document.planet"/> is <value
                expr="document.position"/>.
        </if>
        <if cond="document.topic == 'orbit'">
            It takes <value expr="document.planet"/> <value
            expr="document.orbit"/> to orbit the sun.
        </if>
```

```
<if cond="document.topic == 'temperature'">
    The average surface temperature of <value
    expr="document.planet"/> is
    <value expr="document.temperature"/>.
</if>
<if cond="document.topic == 'size'">
    <value expr="document.planet"/> is the <value
    expr="document.size"/> planet.
</if>
<if cond="document.topic == 'atmosphere'">
    The atmosphere of <value expr="document.planet"/>
    contains <value expr="document.atmosphere"/>.
</if>
<if cond="document.topic == 'moons'">
    <value expr="document.planet"/> has <value
    expr="document.moons"/>.
</if>
<if cond="document.topic == 'all'">
    At an average distance of <value
    expr="document.distance"/>,
    <value expr="document.planet"/> is <value
    expr="document.position"/>,
    taking <value expr="document.orbit"/> to complete its
    orbit. It has an average surface temperature of <value
    expr="document.temperature"/>. The atmosphere is <value
    expr="document.atmosphere"/>. <value
    expr="document.planet"/> is the <value
    expr="document.size"/> planet, and has <value
    expr="document.moons"/>.
    <goto next="#helloworlds"/>
</if>
<goto next="#whatnext"/>
</block>
</form>
```

Using Breaks to Fine-Tune Timing

You can use <break> tags to fine-tune the timing of pauses in a user interface. For example, if a menu has more than three options, it's advantageous to separate the options with 750-msec pauses. These pauses are long enough to invite users to barge in to make a selection, but are short enough that they do not adversely affect the time required to present all of the options. Another use for <break> is to define the length of the pause that follows a nondirective prompt before presenting a set of specific options (or, if there are too many options, playing an example help), typically set to 2–3 seconds. If a menu occurs at a task-terminal point, then studies of videotaped usability sessions suggest that it's a good idea to present the set of always-active navigation commands about 1500–2000 ms after the end of the primary menu. The following lines of code illustrate all three of these uses for <break>.

> Which planet are you interested in? <break time="3000ms"/>
> Select Mercury, <break time="750ms"/>
>> Venus, <break time="750ms"/>
>> Earth, <break time="750ms"/>
>> Mars, <break time="750ms"/>
>> Jupiter, <break time="750ms"/>
>> Saturn, <break time="750ms"/>
>> Uranus, <break time="750ms"/>
>> Neptune, <break time="750ms"/>
>> Or Pluto.
> <break time="1500ms"/> You can always say Repeat, Go Back, Start Over, or Exit.

Another important use of the <break> tag in prototypes using TTS is to place a small pause (150–250 ms) at the beginning of system prompts and messages. The purpose of this small pause is to help users who have barged into a prompt or message to detect the change between the message they have interrupted and the next message. Without this break, the audio for the messages can run together, making it difficult to tell when one has stopped and the next has started. For example:

> <break time="150ms"/>At any time you can say Repeat, Help, Go Back, Start Over, or Exit. To continue, say Mercury, Venus, Earth, Mars, Jupiter, Saturn, Uranus, Neptune, or Pluto.

```
<?xml version="1.0" encoding="iso-8859-1"?>
<!DOCTYPE vxml PUBLIC "-//W3C//DTD VOICEXML 2.0//EN"
   "vxml20-1115.dtd">
<vxml version="2.0" xmlns="http://www.w3.org/2001/vxml">
<meta name="GENERATOR" content="Voice Toolkit for Websphere
   Studio"/>

<property name="bargeintype" value="speech"/>

<var name="skipintro" expr="'play'"/>
<var name="helpcounter" expr="0"/>
<var name="distance"/>
<var name="position"/>
<var name="orbit"/>
<var name="temperature"/>
<var name="size"/>
<var name="atmosphere"/>
<var name="moons"/>
<var name="goback"/>
<var name="distance"/>
<var name="currentform"/>

<link next="#helloworlds">
   <grammar version="1.0" mode="voice" root="returntomain">
      <rule id="returntomain" scope="public">
         <one-of>
            <item>main menu</item>
            <item>start over</item>
         </one-of>
      </rule>
   </grammar>
</link>
```

continued

FIGURE 9.4
Hello Worlds Version 4.

```
<link next="#confirmexit">
   <grammar version="1.0" mode="voice" root="goodbye">
      <rule id="goodbye" scope="public">
         <one-of>
             <item>goodbye</item>
             <item>exit</item>
         </one-of>
      </rule>
   </grammar>
</link>

<link next="#goback">
   <grammar version="1.0" mode="voice" root="return">
      <rule id="return" scope="public">
         <one-of>
             <item>go back</item>
         </one-of>
      </rule>
   </grammar>
</link>

<form id="helloworlds">
   <block name='introduction'>
      <assign name="helpcounter" expr="0"/>
      <assign name="currentform" expr="'helloworlds'"/>
      <if cond="skipintro == 'skip'"/>
         <goto nextitem="planet"/>
      </if>
         Welcome to Hello Worlds! Your voice site for information
            about the planets of the solar system. You can say help or
            repeat at any time.
   </block>
```

continued

FIGURE 9.4 (continued)
Hello Worlds Version 4.

```
<field name='planet'>
    <prompt>
        <break time="150ms"/>
            Which planet are you interested in? <break time="3s"/>
            Select Mercury, <break time="750ms"/>
            Venus, <break time="750ms"/>
            Earth, <break time="750ms"/>
            Mars, <break time="750ms"/>
            Jupiter, <break time="750ms"/>
            Saturn, <break time="750ms"/>
            Uranus, <break time="750ms"/>
            Neptune, <break time="750ms"/>
            Pluto, <break time="750ms"/>
            or Exit.
    </prompt>
    <catch event="help noinput nomatch">
        <assign name="helpcounter" expr="helpcounter+1"/>
        <if cond="helpcounter == 1">
            <prompt>
                <break time="150ms"/>Please say the name of a
                    planet.
                <break time="2s"/> Select Mercury, Venus, Earth,
                    Mars, Jupiter, Saturn, Uranus, Neptune, or Pluto.
            </prompt>
        </if>
        <if cond="helpcounter == 2">
            <prompt>
                <break time="150ms"/>At any time you can say
                    Repeat, Help, Go Back, Start Over, or Exit. To con-
                    tinue, say Mercury, Venus, Earth, Mars, Jupiter,
                    Saturn, Uranus, Neptune, or Pluto.
            </prompt>
                <assign name="helpcounter" expr="0"/>
        </if>
    </catch>
```

continued

FIGURE 9.4 (continued)
Hello Worlds Version 4.

```
<grammar version="1.0" mode="voice" root="planet">
    <rule id="planet" scope="public">
        <one-of>
            <item>mercury</item>
            <item>venus</item>
            <item>earth</item>
            <item>mars</item>
            <item>jupiter</item>
            <item>saturn</item>
            <item>uranus</item>
            <item>neptune</item>
            <item>pluto</item>
        </one-of>
    </rule>
</grammar>
<filled>
    <assign name="document.planet" expr="planet"/>
    <assign name="document.skipintro" expr="'skip'"/>
    <goto next="#playlandmark"/>
</filled>
    </field>
</form>

<form id="playlandmark">
    <block>
        <audio src="triple.au"/>
        <prompt>
            <break time="150ms"/>
            <value expr="document.planet"/>
            <break time="150ms"/>
            <audio src="triple.au"/>
        </prompt>
```

continued

FIGURE 9.4 (continued)
Hello Worlds Version 4.

```
        <goto next="#gettopic"/>
    </block>
</form>

<form id='gettopic'>
    <block>
        <assign name="helpcounter" expr="0"/>
        <assign name="document.goback" expr="'helloworlds'"/>
        <assign name="currentform" expr="'gettopic'"/>
    </block>
    <field name="topic">
        <prompt>
            <break time="150ms"/>Select a topic. <break time="3s"/>
            Select Distance from Sun, <break time="750ms"/>
            Orbital Period, <break time="750ms"/>
            Temperature, <break time="750ms"/>
            Size, <break time="750ms"/>
            Atmospheric Composition, <break time="750ms"/>
            Number of Moons, <break time="750ms"/>
            or Tell Me Everything.
        </prompt>
        <catch event="help noinput match">
            <assign name="helpcounter" expr="helpcounter+1"/>
            <if cond="helpcounter == 1">
                <prompt>
                    <break time="150ms"/>Please say the topic you're
                        interested in. Select Distance from Sun, Orbital
                        Period, Temperature, Size, Atmospheric
                        Composition, Number of Moons, or Tell Me
                        Everything.
                </prompt>
            </if>
```

continued

FIGURE 9.4 (continued)
Hello Worlds Version 4.

```
<if cond="helpcounter == 2">
    <prompt>
        <break time="150ms"/>At any time you can say
            Repeat, Help, Go Back, Start Over, or Exit. To
            continue, say Distance from Sun, Oribital Period,
            Temperature, Size, Atmospheric Composition,
            Number of Moons, or Tell Me Everything.
    </prompt>
    <assign name="helpcounter" expr="0"/>
</if>
</catch>
<grammar version="1.0" mode="voice" root="planetproperty">
    <rule id="planetproperty" scope="public">
        <one-of>
            <item>distance from sun<tag>$="distance"</tag>
                </item>
            <item>orbital period<tag>$="orbit"</tag></item>
            <item>temperature</item>
            <item>size</item>
            <item>atmosphericcomposition<tag>$="atmosphere"
                </tag></item>
            <item>number of moons<tag>$="moons"</tag>
                </item>
            <item>tell me everything<tag>$="all"</tag></item>
            <item>no<tag>$="goback"</tag></item>
            <item>thats not right<tag>$="goback"</tag></item>
        </one-of>
    </rule>
</grammar>
<filled>
    <assign name="document.topic" expr="topic"/>
    <if cond="topic=='goback'">
        <goto next="#helloworlds"/>
    </if>
```

continued

FIGURE 9.4 (continued)
Hello Worlds Version 4.

```
            <goto next="#lookupdata"/>
        </filled>
    </field>
</form>

<form id='lookupdata'>
    <block>
    <prompt>
    </prompt>
        <if cond="document.planet == 'mercury'">
            <assign   name="document.distance"   expr="'58   million
                kilometers'"/>
            <assign name="document.position" expr="'the closest planet
                to the sun'"/>
            <assign name="document.orbit" expr="'88 days'"/>
            <assign name="document.temperature" expr="'440 degrees
                Celcius'"/>
            <assign name="document.size" expr="'next to smallest'"/>
            <assign name="document.atmosphere" expr="'98% helium,
                2% hydrogen'"/>
            <assign name="document.moons" expr="'no moons'"/>
        </if>
        <if cond="document.planet == 'venus'">
            <assign   name="document.distance"   expr="'108   million
                kilometers'"/>
            <assign   name="document.position"   expr="'the   second
                planet from the sun'"/>
            <assign name="document.orbit" expr="'224 days'"/>
            <assign name="document.temperature" expr="'457 degrees
                Celcius, the hottest in the solar system due to a runaway
                greenhouse effect'"/>
            <assign name="document.size" expr="'sixth largest'"/>
            <assign  name="document.atmosphere"  expr="'97%  carbon
                dioxide, 3% nitrogen'"/>
```

continued

FIGURE 9.4 (continued)
Hello Worlds Version 4.

```
        <assign name="document.moons" expr="'no moons'"/>
</if>
<if cond="document.planet == 'earth'">
    <assign  name="document.distance"  expr="'150  million
        kilometers'"/>
    <assign name="document.position" expr="'the third planet
        from the sun'"/>
    <assign name="document.orbit" expr="'365 days'"/>
    <assign name="document.temperature" expr="'15 degrees
        Celcius'"/>
    <assign name="document.size" expr="'fifth largest'"/>
    <assign name="document.atmosphere" expr="'79% nitrogen,
        21% oxygen'"/>
    <assign name="document.moons" expr="'one moon'"/>
</if>
<if cond="document.planet == 'mars'">
    <assign  name="document.distance"  expr="'228  million
        kilometers'"/>
    <assign name="document.position" expr="'the fourth planet
        from the sun'"/>
    <assign name="document.orbit" expr="'687 days'"/>
    <assign name="document.temperature" expr="'negative 55
        degrees Celcius'"/>
    <assign name="document.size" expr="'seventh largest'"/>
    <assign name="document.atmosphere" expr="'96% carbon
        dioxide, 3% nitrogen, 1% argon'"/>
    <assign name="document.moons" expr="'two moons'"/>
</if>
<if cond="document.planet == 'jupiter'">
    <assign  name="document.distance"  expr="'778  million
        kilometers'"/>
    <assign name="document.position" expr="'the fifth planet
        from the sun'"/>
    <assign name="document.orbit" expr="'12 years'"/>
```

continued

FIGURE 9.4 (continued)
Hello Worlds Version 4.

```
            <assign name="document.temperature" expr="'negative 153
                degrees Celcius'"/>
            <assign name="document.size" expr="'largest'"/>
            <assign name="document.atmosphere" expr="'90% hydrogen,
                10% helium'"/>
            <assign name="document.moons" expr="'39 moons'"/>
        </if>
        <if cond="document.planet == 'saturn'">
            <assign   name="document.distance"   expr="'1.4   billion
                kilometers'"/>
            <assign name="document.position" expr="'the sixth planet
                from the sun'"/>
            <assign name="document.orbit" expr="'29 and a half years'"/>
            <assign name="document.temperature" expr="'negative 185
                degrees Celcius'"/>
            <assign name="document.size" expr="'next to largest'"/>
            <assign name="document.atmosphere" expr="'75% hydrogen,
                25% helium'"/>
            <assign name="document.moons" expr="'30 moons'"/>
        </if>
        <if cond="document.planet == 'uranus'">
            <assign   name="document.distance"   expr="'2.9   billion
                kilometers'"/>
            <assign   name="document.position"   expr="'the   seventh
                planet from the sun'"/>
            <assign name="document.orbit" expr="'84 years'"/>
            <assign name="document.temperature" expr="'negative 215
                degrees Celcius'"/>
            <assign name="document.size" expr="'third largest'"/>
            <assign name="document.atmosphere" expr="'83% hydrogen,
                15% helium, 2% methane'"/>
            <assign name="document.moons" expr="'21 moons'"/>
        </if>
```

continued

FIGURE 9.4 (continued)
Hello Worlds Version 4.

```
        <if cond="document.planet == 'neptune'">
            <assign    name="document.distance"    expr="'4.5    billion
                kilometers'"/>
            <assign name="document.position" expr="'usually the eighth
                planet from the sun'"/>
            <assign name="document.orbit" expr="'165 years'"/>
            <assign name="document.temperature" expr="'negative 225
                degrees Celcius'"/>
            <assign name="document.size" expr="'fourth largest'"/>
            <assign name="document.atmosphere" expr="'85% hydrogen,
                13% helium, 2% methane'"/>
            <assign name="document.moons" expr="'eight moons'"/>
        </if>
        <if cond="document.planet == 'pluto'">
            <assign    name="document.distance"    expr="'5.9    billion
                kilometers'"/>
            <assign name="document.position" expr="'usually the farthest
                planet from the sun'"/>
            <assign name="document.orbit" expr="'248 and a half years'"/>
            <assign name="document.temperature" expr="'negative 233
                degrees Celcius, only 40 degrees above absolute zero'"/>
            <assign name="document.size" expr="'smallest'"/>
            <assign name="document.atmosphere" expr="'mostly nitro-
                gen, with some carbon monoxide and methane'"/>
            <assign name="document.moons" expr="'one moon'"/>
        </if>
        <goto next="#playit"/>
    </block>
</form>

<form id='playit'>
    <block>
        <if cond="document.topic == 'distance'">
```

continued

FIGURE 9.4 (continued)
Hello Worlds Version 4.

```
      At an average distance of
      <value expr="document.distance+', '"/>
      <value expr="document.planet"/> is
      <value expr="document.position+'.'"/>
  </if>
  <if cond="document.topic == 'orbit'">
      It takes <value expr="document.planet"/>
      <value expr="document.orbit"/> to orbit the sun.
  </if>
  <if cond="document.topic == 'temperature'">
      The average surface temperature of
      <value expr="document.planet"/> is
      <value expr="document.temperature+'.'"/>
  </if>
  <if cond="document.topic == 'size'">
      <value expr="document.planet"/> is the
      <value expr="document.size"/> planet.
  </if>
  <if cond="document.topic == 'atmosphere'">
      The atmosphere of <value expr="document.planet"/>
      contains <value expr="document.atmosphere+'.'"/>
  </if>
  <if cond="document.topic == 'moons'">
      <value expr="document.planet"/> has
      <value expr="document.moons+'.'"/>
  </if>
  <if cond="document.topic == 'all'">
      At an average distance of
      <value expr="document.distance+', '"/>
      <value expr="document.planet"/> is
      <value expr="document.position+', '"/>
```

continued

FIGURE 9.4 (continued)
Hello Worlds Version 4.

```
            taking <value expr="document.orbit"/>
            to complete its orbit.
            It has an average surface temperature of
            <value expr="document.temperature+'.'"/>
            The atmosphere is <value expr="document.atmosphere+'.'"/>
            <value expr="document.planet"/> is the
            <value expr="document.size"/> planet, and has
            <value expr="document.moons+'. '"/>
            <goto next="#helloworlds"/>
        </if>
        <goto next="#whatnext"/>
    </block>
</form>

<form id='whatnext'>
    <block>
        <assign name="helpcounter" expr="0"/>
        <assign name="document.goback" expr="'gettopic'"/>
        <assign name="currentform" expr="'whatnext'"/>
    </block>
    <block>
        <prompt>
            <break time="150ms"/>
            Select another topic or another planet.
            <break time="2s"/>
            For   more   information   about   <value   expr="document.
                planet"/>,
        </prompt>
    </block>
    <field name="topic2">
        <prompt>
            select Distance from Sun, <break time="750ms"/>
```

continued

FIGURE 9.4 (continued)
Hello Worlds Version 4.

```
            Orbital Period, <break time="750ms"/>
            Temperature, <break time="750ms"/>
            Size, <break time="750ms"/>
            Atmospheric Composition, <break time="750ms"/>
            Number of Moons, <break time="750ms"/>
            or Tell Me Everything.
      </prompt>
<catch event="help noinput nomatch">
<assign name="helpcounter" expr="helpcounter+1"/>
<if cond="helpcounter == 1">
    <prompt>
        <break time="150ms"/>Please say the topic you're interested
            in. Select Distance from Sun, Orbital Period, Temperature,
            Size, Atmospheric Composition, Number of Moons, or
            Tell Me Everything.
    </prompt>
</if>
<if cond="helpcounter == 2">
    <prompt>
        <break time="150ms"/>At any time you can say Repeat,
            Help, Go Back, Start Over, or Exit. To continue, select
            Distance from Sun, Orbital Period, Temperature, Size,
            Atmospheric Composition, Number of Moons, or Tell Me
            Everything.
    </prompt>
    <assign name="helpcounter" expr="0"/>
</if>
</catch>
<grammar version="1.0" mode="voice" root="planetprop2">
    <rule id="planetprop2" scope="public">
        <one-of>
            <item>distance from sun<tag>$="distance"</tag></item>
            <item>orbital period<tag>$="orbit"</tag></item>
```

continued

FIGURE 9.4 (continued)
Hello Worlds Version 4.

```
            <item>temperature</item>
            <item>size</item>
            <item>atmospheric    composition<tag>$="atmosphere"
                </tag></item>
            <item>number of moons<tag>$="moons"</tag></item>
            <item>tell me everything<tag>$="all"</tag></item>
            <item>repeat</item>
            <item>mercury</item>
            <item>venus</item>
            <item>earth</item>
            <item>mars</item>
            <item>jupiter</item>
            <item>saturn</item>
            <item>uranus</item>
            <item>neptune</item>
            <item>pluto</item>
            <item>select   another   topic<tag>$="newtopic"</tag>
                </item>
            <item>select  another  planet<tag>$="newplanet"</tag>
                </item>
        </one-of>
    </rule>
</grammar>
<filled>
    <if cond="topic2 == 'newplanet'">
        <goto next="#helloworlds"/>
    </if>
    <if cond="topic2 == 'newtopic'">
        <goto nextitem="topic2"/>
    </if>
    <if cond="topic2 == 'mercury'">
        <assign name="document.planet" expr="'mercury'"/>
```

continued

FIGURE 9.4 (continued)
Hello Worlds Version 4.

```
      <goto next="#playlandmark"/>
   </if>
   <if cond="topic2 == 'venus'">
      <assign name="document.planet" expr="'venus'"/>
      <goto next="#playlandmark"/>
   </if>
   <if cond="topic2 == 'earth'">
      <assign name="document.planet" expr="'earth'"/>
      <goto next="#playlandmark"/>
   </if>
   <if cond="topic2 == 'mars'">
      <assign name="document.planet" expr="'mars'"/>
      <goto next="#playlandmark"/>
   </if>
   <if cond="topic2 == 'jupiter'">
      <assign name="document.planet" expr="'jupiter'"/>
      <goto next="#playlandmark"/>
   </if>
   <if cond="topic2 == 'saturn'">
      <assign name="document.planet" expr="'saturn'"/>
      <goto next="#playlandmark"/>
   </if>
   <if cond="topic2 == 'uranus'">
      <assign name="document.planet" expr="'uranus'"/>
      <goto next="#playlandmark"/>
   </if>
   <if cond="topic2 == 'neptune'">
      <assign name="document.planet" expr="'neptune'"/>
      <goto next="#playlandmark"/>
   </if>
   <if cond="topic2 == 'pluto'">
```

continued

FIGURE 9.4 (continued)
Hello Worlds Version 4.

```
            <assign name="document.planet" expr="'pluto'"/>
            <goto next="#playlandmark"/>
        </if>
        <if cond="topic2 == 'repeat'">
            <goto next="#playit"/>
        </if>
        <assign name="document.topic" expr="topic2"/>
        <goto next="#lookupdata"/>
        </filled>
    </field>
</form>

<form id='goback'>
    <block>
        <goto expr="'#'+document.goback"/>
    </block>
</form>

<form id='confirmexit'>
    <field name="exitChoice" type="boolean">
        <prompt>
            <break time="150ms"/>
            Do you want to end this call?
        </prompt>
        <catch event="help noinput nomatch">
        <assign name="helpcounter" expr="helpcounter+1"/>
        <if cond="helpcounter == 1">
            <prompt>
                <break time="150ms"/>Please say Yes, No, or Repeat.
            </prompt>
        </if>
```

continued

FIGURE 9.4 (continued)
Hello Worlds Version 4.

```
        <if cond="helpcounter == 2">
            <prompt>
                <break time="150ms"/>At any time you can say Repeat,
                    Help, Go Back, Start Over, or Exit. To end the call, say
                    Yes. To return to Hello Worlds, say No.
            </prompt>
            <assign name="helpcounter" expr="0"/>
        </if>
        </catch>
        <filled>
            <if cond="exitChoice">
                <goto next="#exit"/>
                <else/>
                <audio src="triple.au"/>
                <prompt>
                    <break time="150ms"/>
                    Returning.
                </prompt>
                <goto expr="'#'+document.currentform"/>
            </if>
        </filled>
    </field>
</form>

<form id='exit'>
    <block>
        <prompt>
            <break time="150ms"/>
            Thanks for calling Hello Worlds. Goodbye!
        </prompt>
        <exit/>
    </block>
</form>

</vxml>
```

FIGURE 9.4 (continued)
Hello Worlds Version 4.

Using Recorded Speech

To enhance the application, you can make a few changes so it will play recorded speech.

Replacing TTS with Recorded Speech

Replacing static TTS with recorded speech is reasonably easy for a prototype, as long as you're willing to record the audio yourself or you have someone who will do it for you. To do the conversion for text initially programmed for production via TTS, you enclose the text in <audio> and </audio> tags, and specify the name (and, if necessary, the required directory path) using the 'src' attribute in the <audio> tag, as shown below. Because any large project will have many audio files, it's a good idea to put the audio files in their own folder in the project (the example shows the path to a folder named 'audio').

<audio src="audio/intro.au">

    Welcome to Hello Worlds! Your voice site for information about the
        planets of the solar system.

</audio>

Use any audio editor to create the file. Most VoiceXML browsers will work with 8 kHz .au files. If you are using a separate directory for the audio files, be sure to save it in the audio directory. If the system can't find the specified audio file, it will use play the enclosed text with TTS.

Editing Recorded Audio Segments

After making your initial recordings, use the audio editor to equalize the volume of your recordings, trim beginning and ending silences, and take care of any required editing of segment-internal pauses (either lengthening or shortening). Figure 9.5 shows the waveforms of an original and trimmed audio segment. As shown in Figure 9.5, almost all of the silence has been trimmed (deleted) from the beginning and end of the segment. Unless there is a specific reason to do otherwise, you should trim recorded speech aggressively to avoid unwanted delays in the playing of audio segments (caused by silence at the beginning of the segment) and unwanted extensions of the silence timeout period (caused by silence at the end of a segment).

Dynamic Selection of Audio Segments

In VoiceXML 2.0, you can dynamically select an audio segment to play (like the dynamic methods available for controlling the data played by TTS). This

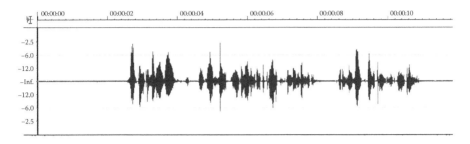

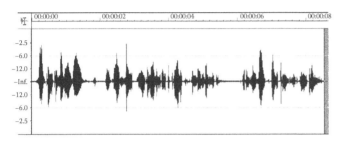

FIGURE 9.5
Original and trimmed audio segments.

feature is available by using 'expr' in place of 'src' inside an <audio> tag, as shown in the fourth line of the following form:

```
<form id="playlandmark">
    <block>
        <audio src="audio/triple.au"/>
            <audio expr="'audio/'+document.planet+'.au'">
            <break time="150ms"/>
            <value expr="document.planet"/>
            <break time="150ms"/>
            </audio>
        <audio src="audio/triple.au"/>
        <goto next="#gettopic"/>
    </block>
</form>
```

Note: If you plan to play two audio segments together to make a larger segment, try to end the first segment at a natural pause point (for example, a period or comma) to help make the concatenation of the segments sound as natural as possible.

Using an Application Root Document

All examples in this chapter have coded the entire application in one document, with variable and grammar scope set to 'document' level as required. An alternative structure uses multiple documents, with an 'application root document' containing always-active commands and other common code. As long as a set of VoiceXML documents refers to the application root document it stays loaded, and its variables and grammars, if set to document scope on the application root document, are available for use on any page of the application. Although I have not yet encountered a need for this, very complex prototypes might be easier to manage if broken up into several documents that refer to a common application root document. References to an application root document appear in the <vxml> element:

```
<vxml version="2.0" application="app-root.vxml">
```

Summary

It is important to read about research and practice in SUI design, but it is also valuable to experience different designs. With the introduction of VoiceXML, it is possible for non-programmers to put together SUI design prototypes for the purpose of expert review, formative usability testing, or experimental comparison of designs. From a basic "Hello, World" to an application that provides information about the planets of the solar system, this chapter provides programming examples for a number of SUI design elements, including:

- Playing an introduction once
- Defining always-active commands with links
- Using grammar tags to classify responses
- Setting variables to document-level scope
- Directing the call flow with if statements in the filled section
- Using the document-level variables and unconditional return to main
- Exiting without confirmation
- Creating self-revealing help
- Defining an unconditional go back
- Defining a conditional go back
- Exiting with confirmation
- Integrating audio tones

- Simulating data acquisition from a backend server
- Playing the data
- Using breaks to fine-tune timing
- Replacing TTS with recorded speech
- Editing recorded audio segments
- Dynamic selection of audio segments
- Using an application root document

References

Lewis, J. R. (2004). *Prototyping best-practice speech user interfaces with VoiceXML 2.0 and the IBM VoiceXML toolkit* (Tech. Rep. 29.3732, available at: drjim.0catch.com/vxml20proto-ral.pdf). Raleigh, NC: IBM Corp.

10

Final Words

I Appreciate Your Patience

Here it is—the end of the book. I hope you found it interesting and useful. Of all of the areas of human factors design and research in which I have had the opportunity to work, SUI design has been the most fascinating and enjoyable.

While reviewing the research, I came across data that I had formerly not known. Some of this research confirmed my current design practices, but more importantly, other research has led me to make some changes in my design strategies, directly as a consequence of researching and writing this book. With any luck, you, too, will have discovered some aspect of SUI research that you can use to improve your designs.

No human artifact is perfect, and this book is no exception, despite my best efforts and the efforts of my reviewers and publisher. As I worked through the research literature I had compiled in the first part of 2009, I sometimes came across lines of research that I had missed in the first pass, and I had to track back to pick up the missing research. There may be other lines of research that I missed completely, but I have strived to make the guidance (both practice- and research-based) provided in this book reasonably solid. To the extent that I have succeeded, that's great. To the extent that I have not, well, I appreciate your patience.

Please Hold for the Next Available Research

Books are static. The world is not. By the time you read this, at a minimum, months will have passed since the submission for publication of the final manuscript of this book. Speech technologies will continue to evolve, and researchers interested in SUI design and related topics will continue to publish their findings.

From the perspective of a SUI designer, I care only a little about the continuing development of speech technologies. For example, it doesn't matter to me whether the underlying technology for speech recognition is hidden Markov models or neural nets. I do care, however, about the operating characteristics of the technologies as they affect the human factors of design. Speech recognition accuracy with modern methods tends to be very high, so I have little expectation of remarkable future improvements in that area. On the other hand, TTS technologies have shown remarkable improvement over the past 10 years, and there could be exciting new artificial voices and voice-development methods in the future. Another research area to watch is the continuing development of intelligent dialog managers, possibly assisted by emerging advanced natural language technologies.

In contrast to my slight interest in most future speech technologies, I cannot wait to see what innovations lie ahead in clever use of speech technologies to solve human needs, both in the near and far term. I encourage researchers from all the fields related to SUI design to put their theories to the test and to share the results through peer-reviewed publication, whether they are consistent or inconsistent with theory. Theories evolve by incorporating new, unexpected data, as should design practices.

There are a number of ways to keep up to date on research related to SUI design:

- Consider joining the new Association for Voice Interaction Design (avixd.org). New research is sure to be a topic of conversation for that organization, and in the future it is possible that they might sponsor a research journal.

- For a broader range of speech topics that includes SUI design consider joining the Applied Voice Input-Output Society (avios.org).

- Get a subscription to *Speech Technology Magazine* (speechtechmag. com), offered free of charge to qualified subscribers in the United States. Although it is not a scientific journal, and therefore does not meet a high level of evidentiary standards, it has for more than a decade provided a venue for thought leaders in the field of SUI design to publish articles about a wide variety of topics.

- Attend trade meetings like SpeechTek (speechtek.com) and professional meetings sponsored by organizations such as AVIOS, the International Speech Communication Association (isca-speech.org), the Human Factors and Ergonomics Society (hfes.org), the ACM special interest group in computer–human interaction (sigchi.org), Human-Computer Interaction International (hci-international.org), the Interact conferences of the International Federation for Information Processing (ifip.org), and the Usability Professionals Association (upassoc.org). If you cannot attend, these professional organizations publish the research papers presented at their meetings.

- Stay current with the research published in peer-reviewed scientific journals. When reading the scientific literature do so critically. It is not wise to just accept the statements made in the abstract. Read the paper, think about its related research to see where it fits in, be aware of the limitations to generalization due to participants and tasks, and judge for yourself whether the researchers have properly analyzed and interpreted the data. Some of the scientific journals that at least occasionally include research on SUI design (and post their tables of contents online), starting with those most focused on the topic, are:

- *International Journal of Speech Technology* (springerlink.com/content/100275/)
- *Speech Communication* (sciencedirect.com/science/journal/01676393)
- *Computer Speech and Language* (sciencedirect.com/science/journal/08852308)
- *Human Factors* (hfes.org/publications)
- *Natural Language Engineering* (journals.cambridge.org/action/displayJournal?jid=NLE)
- *ACM Transactions on Speech and Language Processing* (portal.acm.org/toc.cfm?id=J957)
- *Computational Linguistics* (mitpressjournals.org/loi/coli)
- *ACM Transactions on Computer-Human Interaction* (portal.acm.org/toc.cfm?id=J756)
- *Behaviour and Information Technology* (informaworld.com/smpp/title~content=t713736316~db=all)
- *Interacting with Computers* (sciencedirect.com/science/journal/09535438)
- *International Journal of Human-Computer Studies* (sciencedirect.com/science/journal/10715819)
- *Human Computer Interaction* (informaworld.com/smpp/title~content=t775653648~db=all)
- *International Journal of Human-Computer Interaction* (informaworld.com/smpp/title~db=all~content=t775653655)
- *Journal of Usability Studies* (upassoc.org/upa_publications/jus/)

Thanks for Reading, Goodbye!

Thank you for reading this book. For all the SUI practitioners, I wish you the best of luck in your future SUI designs. For business analysts and project managers, I hope you now have a better understanding and appreciation of what your designers do, and how to avoid some common but ineffective design practices.

Finally, for graduate students and others looking for research topics in SUI design, it should be clear that there are many interesting areas that need further investigation to provide better guidance to SUI design, with results either confirming or disconfirming current design practice. SUI practitioners are counting on you, and I personally am waiting with great interest to see what you discover.

Index